D1346079

Quark Confinement and Field Theory

Quark Confinement and Field Theory

Proceedings of a Conference at
The University of Rochester
Rochester, New York
June 14-18, 1976

Edited by

D. R. STUMP

D. H. WEINGARTEN

A Wiley-Interscience Publication

JOHN WILEY & SONS

New York / London / Sydney / Toronto

Library of Congress Cataloging in Publication Data:

Quark confinement and field theory.

"A Wiley-Interscience publication."

1. Quarks—Congresses. 2. Field theory (Physics)—Congresses. I. Stump, Daniel R. II. Weingarten, Don H. III. Rochester, N. Y. University.

QC793.5.Q252Q36 539.7′216 76-39963
ISBN 0-471-02721-9

Printed in the United States of America

10 9 8 7 6 5 4 3 2 1

Preface

During the week of June 14, 1976 a workshop on quark binding was held at the University of Rochester. Those in attendance included fifteen invited speakers, a few visitors, and faculty and graduate students of the Department of Physics and Astronomy of the University of Rochester. The title, "Quark Binding," was chosen only as a catchall phrase with no attempt being made to restrict the speakers' topics (indeed, the range of topics was rather wide).

The workshop was intended to provide an opportunity for participants to describe the results of current research on one of the most active and intriguing frontiers of particle physics. Therefore, considerable effort has been made to achieve publication of these proceedings as quickly as possible. Heartfelt thanks are due to the authors for their cooperation in this effort by their prompt submission of manuscripts. Acknowledgment must also be made of the assistance provided by graduate students of the High Energy Theory Group, who carefully proofread the final versions of the manuscript.

We are pleased to thank the U.S.Energy Research and Development Administration for financial support of the conference. Finally, we wish to express our appreciation to Edna Hughes, whose prompt typing of the manuscript was indispensable to the publication of these proceedings.

Daniel Stump

Don Weingarten

Rochester, New York

Bloomington, Indiana

November 1976

Contents

Quark Confinement and Field Theory

Strings, Vortices, and Gauge Fields

YOICHIRO NAMBU
University of Chicago
Chicago, Illinois

In this talk I will make a few scattered observations on mathematical formalisms which are aimed at describing stringlike objects and their interactions. The attempt was originally motivated by the string model of hadrons, which may well be a phenomenological manifestation of a gauge theory of strong interactions. Here, however, I will put emphasis on broader aspects of the problem, keeping in mind that there may exist as yet unknown phenomena which are to be described by these mathematical formalisms. Of course it is sheer speculation at the moment.

There are three main topics to be discussed. They are: relativistic hydrodynamics of vortices, its broken symmetry schemes and non-Abelian versions, and a unified algebraic characterization of string and gauge field equations.

I. RELATIVISTIC HYDRODYNAMICS OF VORTICES

At the outset I would like to make clear that this is an elaboration of the ideas developed originally by Kalb and Ramond[1]. The emphasis is here on its physical interpretation and implications. First let us define the following fields.

Velocity field 4-vector (dimension M^2)	V_μ
Velocity potential 6-vector (dimension M)	$W_{\mu\nu}$ $(=-W_{\nu\mu})$
Vorticity 6-vector (dimension M^3)	$\Omega_{\mu\nu}$ $(=-\Omega_{\nu\mu})$
Vorticity source 6-vector (dimension M^2)	$\omega_{\mu\nu}$ $(=-\omega_{\nu\mu})$

(1)

V_μ and $\Omega_{\mu\nu}$ are derived from $W_{\mu\nu}$ according to

$$
\begin{aligned}
V_\lambda &\equiv i\varepsilon_{\lambda\mu\nu\rho}\,\partial_\mu W_{\nu\rho} \equiv \partial_\mu \tilde{W}_{\mu\lambda}, \\
\partial_\lambda V_\lambda &\equiv 0 \\
\Omega_{\mu\nu} &\equiv \varepsilon_{\mu\nu\lambda\rho}\partial_\lambda V_\rho \quad (\text{or } \tilde{\Omega}_{\mu\nu} = -\partial_\mu V_\nu + \partial_\nu V_\mu) \\
\partial_\mu \Omega_{\mu\nu} &\equiv 0 \, .
\end{aligned} \tag{2}
$$

The source $\omega_{\mu\nu}$ also is assumed to satisfy the continuity equation

$$\partial_\mu \omega_{\mu\nu} = 0 \, . \tag{3}$$

The velocity field V_μ is a physical observable; actually it should be interpreted as a current density, namely density × 4-velocity. The velocity potential $W_{\mu\nu}$, on the other hand, is not uniquely determined by V_μ, but is subject to a gauge transformation

$$W_{\mu\nu} \to W_{\mu\nu} + \partial_\mu \Lambda_\nu - \partial_\nu \Lambda_\mu \, , \tag{4}$$

with an arbitrary vector field Λ_μ. $W_{\mu\nu}$, $\Omega_{\mu\nu}$ and $\omega_{\mu\nu}$ are defined in such a way that in the nonrelativistic limit the space-time (or "electric") components are nonzero and correspond to the usual definition in nonrelativistic hydrodynamics. Note also that in comparison with electrodynamics, the roles of a 4-vector and a 6-vector are reversed, but otherwise we are describing rather similar phenomena. This becomes clearer when we look at the Ramond Lagrangian

$$L = \frac{1}{2} V_\mu V_\mu - \frac{k}{2} W_{\mu\nu}\omega_{\mu\nu} \, , \tag{5}$$

which satisfies gauge invariance (4) because of Eq. (3). k is a parameter with the dimensions of a mass. By varying $W_{\mu\nu}$, we get from this

$$\partial_\mu V_\nu - \partial_\nu V_\mu = k\,\tilde{\omega}_{\mu\nu} \, , \tag{6a}$$

or

$$\Box\, W_{\mu\nu} - (\partial_\mu U_\nu - \partial_\nu U_\mu) = k\,\omega_{\mu\nu} \, . \tag{6b}$$

Here we have introduced another auxiliary field

$$U_\mu \equiv \partial_\lambda W_{\lambda\mu} \, . \tag{7}$$

Equations (6a) and (6b) amount to an identity involving W, U and V. From Eq. (6a) follows

$$\Box V_\nu = \partial_\mu \tilde{\omega}_{\mu\nu} , \tag{8a}$$

and with a special gauge $U_\mu = 0$, $W_{\mu\nu}$ also satisfies

$$\Box W_{\mu\nu} = k\, \omega_{\mu\nu} . \tag{8b}$$

The content of these equations becomes more tranparent if we write out the components of $W_{\mu\nu}$ and $\omega_{\mu\nu}$ as

$$(W_{ij}, W_{io}) = (\varepsilon_{ijk}\, b_k, e_i) ,$$
$$(\tilde{W}_{ij}, \tilde{W}_{io}) = (-\varepsilon_{ijk}\, e_k, b_i) ,$$
$$(\omega_{ij}, \omega_{io}) = (\varepsilon_{ijk}\, \nu_k, \omega_i) , \tag{9}$$

where $\vec{e}$ and $\vec{b}$ are analogs of the electric and magnetic fields. Then

$$(V_i, V_o) = (-\vec{\nabla} \times \vec{e} + \vec{b}, + \vec{\nabla} \cdot \vec{b})$$
$$L = \frac{1}{2} (\vec{b} + \vec{\nabla} \times \vec{e})^2 - \frac{1}{2} (\vec{\nabla} \cdot \vec{b})^2 - k(\vec{\nu} \cdot \vec{b} - \vec{e} \cdot \vec{\omega}), \tag{10}$$

$$\vec{b} - \vec{\nabla}(\vec{\nabla} \cdot \vec{b}) + \vec{\nabla} \times \vec{e} = -k\vec{\nu}$$
$$\vec{\nabla} \times (\vec{b} + \vec{\nabla} \times \vec{e}) = - k\vec{\omega} . \tag{11}$$

From this it is easy to see that 1) $\vec{e}$ is nondynamical (like the Coulomb field in electrodynamics); 2) $\vec{b} + \vec{\nabla} \times \vec{e}$ is longitudinal; and 3) the dynamical field $\vec{b}$ describes massless longitudinal radiation (and quanta).

The emergence of a massless radiation associated with hydrodynamic motion is characteristic of a relativistic field theory, as in the case of electromagnetism. The importance of this observation, however, lies in the implied possibility that such hydrodynamic media with attendant massless, longitudinal radiation (propagating with the speed of light!) might actually exist in nature. What will be the source of such radiation?

The Dirac construction of the source field $\omega_{\mu\nu}$ is well known. The latter represents the surface element of a world sheet swept out by a closed string, which in our interpretation constitutes the core of a vortex ring. Thus

$$\omega_{\mu\nu}(x) = \int \delta^4(x-y)\, d\tau d\sigma\, \sigma_{\mu\nu}(y)$$

$$\sigma_{\mu\nu}(y) = \dot{y}_\mu y'_\nu - y'_\mu \dot{y}_\nu \equiv [y_\mu, y_\nu] (\dot{y}_\mu \equiv \partial y_\mu/\partial\tau, y'_\mu \equiv \partial y_\mu/\partial\sigma). \tag{12}$$

Here the $y_\mu(\tau,\sigma)$ are the coordinates of a point on the sheet, which is parametrized by the internal coordinates τ and σ. The notation is familiar except for the Poisson bracket symbol $[y_\mu, y_\nu]$ which is designed to emphasize the fact that τ and σ act as if they were a pair of canonical variables. They range from $-\infty$ to ∞, but y_μ is assumed to be periodic in σ, so that σ parametrizes a closed string at a given instant τ. If a segment of the string is at rest, and τ is chosen to be proportional to the time coordinate, the only nonzero components $\omega_{oi} = \omega_i$ of $\omega_{\mu\nu}$ represent a tangential vector to the string. Equation (8) then gives rise to a long-range potential created by the string element, which is similar to the vector potential due to an electric current element. The string-string interaction arising from such a potential, however, is of opposite sign to that for the current-current interaction because ω_i is "electric" rather than "magnetic". This is exactly the property of the interaction between hydrodynamic vortex elements.

When a vortex line element parallel to the x axis vibrates in the y direction, it radiates a longitudinal (compressional or "acoustic") wave, with a $\cos^2\theta$ intensity distribution relative to the z axis. This is the new effect mentioned above. But creation and detection of such waves depend on the existence (in vacuo) of vortices. So the next question is: do the quanta of vortex rings exist as elementary particles? A corollary to this question is: does a conserved source field $\omega_{\mu\nu}(x)$ exist in local field theory? And if so, is it associated with any symmetry principle?

There is a trivial answer to the second question. We can put

$$\omega_{\mu\nu} \propto \tilde{F}_{\mu\nu} , \tag{13}$$

where $F_{\mu\nu}$ is an analog of the familiar electromagnetic field, which in turn is generated by a 4-vector current. Or else we can simply write

$$\tilde{\omega}_{\mu\nu} = \partial_\mu j_\nu - \partial_\nu j_\mu , \tag{14}$$

with an arbitrary vector field j_μ. But it is unlikely that any symmetry principle can be associated with these currents. More attractive is the spirit of the original string model, namely to quantize the vortex as a nonlocal object. To each closed vortex, we then associate a wave function $\psi[y(\sigma)]$, which is a functional of the string configuration embedded in the Minkowski space. As was proposed by Ramond, we can define a covariant derivative

$$D_\mu(y) = y'_\nu \left(\frac{\delta}{\delta\sigma_{\mu\nu}(y)} - ig\, W_{\mu\nu}(y) \right) , \qquad (15)$$

where the first term represents the change induced when the vortex line at point y is displaced to create a 2-surface $\delta\sigma_{\mu\nu}$. The second term is designed to absorb the effect of the gauge transformation (4) on $W_{\mu\nu}$ into a phase transformation on Ψ:

$$\Psi[y] \rightarrow \exp[i \int_s \Lambda_\mu \, d\sigma_\mu] \, \Psi[y] \; . \qquad (16)$$

Unfortunately, there is a formidable mathematical problem in choosing an action integral and defining a proper measure in the space of vortex configurations. The well known difficulties encountered in the dual string model are reflections of this problem.

It appears that vortices as elementary particles do not exist;if relativistic vortices exist at all, they are probably composite, nonlocal objects with very large internal degrees of freedom. And the question remains whether long-range hydrodynamic waves can be emitted and absorbed by such objects.

A concrete example of relativistic vortex derived from local field theory as a nonlocal object is the Nielsen-Olesen string.[2] However, it is not coupled to a long-range field, and hence not an analog of hydrodynamic vortex line. Rather, it is a magnetic flux line in a superconductive medium.

II. BROKEN SYMMETRY SCHEMES AND OTHER GENERALIZATIONS

We will discuss here various generalizations of the basic Lagrangian we introduced above (Eq. (5)). These seem to be more interesting and relevant as providing models of hadron confinement.

A. Massive 6-Vector Field

We add a mass term to the Lagrangian (5):

$$L = \frac{1}{2} V_\mu V_\mu - \frac{m^2}{4} W_{\mu\nu} W_{\mu\nu} - k\, W_{\mu\nu} \omega_{\mu\nu} \; , \qquad (17)$$

in analogy to the case of a massive vector field. From Eq. (17) follows

$$\partial_\mu V_\nu - \partial_\nu V_\mu - m^2 \tilde{W}_{\mu\nu} = k\, \tilde{\omega}_{\mu\nu} , \qquad (18)$$

so that

$$(\Box - m^2) V_\nu = k\, \partial_\mu \tilde{\omega}_{\mu\nu} , \qquad (19a)$$

and

$$-m^2 \partial_\mu W_{\mu\nu} = -m^2 U_\nu = k\, \partial_\mu \omega_{\mu\nu} . \qquad (19b)$$

V_ν now becomes a massive spin 1 field. This is nothing but an unorthodox way of describing a vector meson. If $\omega_{\mu\nu}$ is conserved, U_ν must be zero, but the scheme allows for the possibility of nonconserved $\omega_{\mu\nu}$. For example, if the vortex line is open ended, we will have

$$\partial_\mu \omega_{\mu\nu} = j_\nu , \qquad (20)$$

where j_ν is the 4-vector current generated by the two end points, one carrying an opposite charge from the other.

The Lagrangian (17) serves as a phenomenological substitute for the Nielsen-Olesen model. The self-energy of the vortex string is proportional to $m^2 \ell$ where ℓ is its length.[4] The end points of the string may be regarded as a pair of quark and antiquark with equal and opposite charges according to Eq. (20). Thus it will serve as a model for mesons, but baryons cannot be formed this way.

B. Confinement with Nonlocal Fields

We bring in an electromagnetic-type field in addition to $W_{\mu\nu}$ and consider in particular

$$L = \frac{1}{2} V_\mu V_\mu - \frac{1}{4} (m\, W_{\mu\nu} + F_{\mu\nu})(m\, W_{\mu\nu} + F_{\mu\nu}) - A_\mu j_\mu . \qquad (21)$$

Here j_μ is a conserved source current; m is a mass parameter inserted for dimensional reasons. The equations following Eq. (21) are

$$\partial_\mu V_\nu - \partial_\nu V_\mu - m(m\, \tilde{W}_{\mu\nu} + F_{\mu\nu}) = 0$$
$$\partial_\mu (m\, W_{\mu\nu} + F_{\mu\nu}) = j_\nu , \qquad (22)$$

which lead to

$$(\Box - m^2)V_\mu = 0 \tag{23a}$$

$$j_\nu = 0 \ . \tag{23b}$$

In other words, the equations are not self-consistent unless the current vanishes. The underlying reason is simple: $F_{\mu\nu}$ can be eliminated from Eq. (21) by the gauge transformation (4) with $\Lambda_\mu = A_\mu/m$ and hence Eq. (23b).

It is not unreasonable to interpret Eq. (23b) as a statement of quark confinement. For this, however, we have to regard m as a dynamical parameter. If m is replaced by a field m(x), the current j_μ can exist only where $m(x) \sim 0$. Conceivably the sources $\omega_{\mu\nu}$ and j_μ coupled respectively to $W_{\mu\nu}$ and A_μ develop expectation values $\langle\omega,\omega\rangle \sim m^2$, $\langle\omega,j\rangle \sim m$ in vacuo, but that in the vicinity of charges they are forced to vanish. Such a mechanism would reproduce the Nielsen-Olesen string with quarks attached to its ends. The corresponding Lagrangian would be

$$L = \frac{1}{2} V_\mu V_\mu - \frac{1}{4} F_{\mu\nu} F_{\mu\nu} - k\, W_{\mu\nu}\omega_{\mu\nu} - A_\mu j_\mu , \tag{24}$$

where the sources are supposed to satisfy Eq. (20) so that the gauge invariance under Eq. (4) holds with Λ_μ replaced by A_μ/k. The problem here is to find a natural (and renormalizable) candidate for $\omega_{\mu\nu}$ in local field theory. One way to simulate this mechanism would be to regard m in Eq. (21) as a Goldstone scalar field with a nonzero value $\langle m(x)\rangle$ in the absence of j_μ.

C. Non-Abelian Analogs

Most of what has been done above can be formally generalized to non-Abelian cases by the replacement of the derivatives ∂_μ and the gauge field $F_{\mu\nu}$ by their covariant versions. In other words, all fields except the gauge potentials A_μ will be regarded as transforming linearly under local gauge transformations. However, the gauge principle of Ramond cannot be reconciled with the Yang-Mills gauge principle. This is because the transformation (4) cannot be properly generalized. To see this, observe that the ansatz

$$W_{\mu\nu} \to W_{\mu\nu} + D_\mu \Lambda_\nu - D_\nu \Lambda_\mu , \tag{24}$$

does not leave $D_\mu \tilde{W}_{\mu\nu}$ invariant unlike the Abelian case, but rather

$$(D_\mu \tilde{W}_{\mu\nu})^a \to (D_\mu \tilde{W}_{\mu\nu})^a - g\, f^{abc} \tilde{F}^b_{\mu\nu} \Lambda^c_\mu , \tag{26}$$

(the superscript refers to the internal spin). The second term vanishes only if $\Lambda_\mu^c = 0$, or det $\tilde{F} = 0$ when $\sum_b F_{\mu\nu}^b f^{abc}$ is regarded as a transformation matrix acting on Λ_μ^c.

In place of Eq. (25), however, there is a residual invariance under a one-parameter family of transformations

$$W_{\mu\nu}^a \to W_{\mu\nu}^a + \alpha\, F_{\mu\nu}^a \ , \qquad (27)$$

with constant α. As a result, the non-Abelian analogs of Eqs. (21) and (22) lead to results similar to Eq. (23), namely

$$D_\mu (D_\mu V_\nu - D_\nu V_\mu) - m^2 V_\nu = 0$$

$$j_\nu(\psi) + j_\nu(W) = 0 \ . \qquad (28)$$

Here j_ν is made up of contributions from the quarks and the W.

Apparently related to the above peculiarities of the non-Abelian cases is the fact that the field $W_{\mu\nu}^a$ has very limited degrees of freedom even when it is coupled only to the gauge field. The equation of motion (cf. Eq. (6a))

$$D_\mu V_\nu - D_\nu V_\mu = 0 \ , \qquad (29)$$

entails, upon taking a covariant divergence of its dual,

$$f^{abc} \tilde{F}_{\mu\nu}^b V_\nu^c = 0 \ , \qquad (30)$$

i.e., $V_\nu = 0$ or det $F = 0$. Thus a consistent theory of non-Abelian massless antisymmetric field does not seem to exist, and again this may have something to do with confinement.

III. ALGEBRAIC CHARACTERIZATION OF STRING AND GAUGE FIELD EQUATIONS

As is well known, the string Lagrangian

$$L = \frac{1}{2}\, g \equiv \frac{1}{2} \sqrt{-\sigma_{\mu\nu}\sigma_{\mu\nu}} = \sqrt{\dot{y}^2 y'^2 - (\dot{y} \cdot y')^2} \ , \qquad (31)$$

can be linearized by going to the Virasoro gauge:

$$\dot{y}^2 + y'^2 = \dot{y} \cdot y' = 0$$

$$L \to \frac{1}{2} (\dot{y}^2 - y'^2) \ . \qquad (32)$$

In the general case, Eq. (31) leads to

$$[\frac{1}{g}\,\sigma_{\mu\nu},\ y_\nu] = [\frac{1}{g}\,[y_\mu,\ y_\nu],\ y_\nu] = 0. \tag{33}$$

Recently Schild[6] has proposed another choice of gauge g^2 = const, with a Lagrangian

$$L' = \frac{1}{4}\,g^2 = -\,\frac{1}{4}\,\sigma_{\mu\nu}\sigma_{\mu\nu}\ . \tag{34}$$

The resulting equations

$$[[y_\mu,\ y_\nu],\ y_\nu] = 0\ . \tag{35}$$

indeed lead to $\dot{g} = g' = 0$, and hence equivalent to Eq. (33). (This can be seen by contracting Eq. (35) with $\dot{y}_\mu$ or y'_μ.) The transition from Eq. (31) to Eq. (34) is very similar to what is often done in point mechanics, which is an essential step in going over to quantum mechanics. Interestingly, the Schild gauge also works in the presence of the field $W_{\mu\nu}$. The Lagrangian

$$L' = \frac{1}{4}\,g^2 + \frac{1}{2}\,\sigma_{\mu\nu}W_{\mu\nu}(y)\ , \tag{36}$$

leads to

$$[[y_\mu,\ y_\nu] + W_{\mu\nu},\ y_\nu] = \frac{1}{2}\,\sigma_{\nu\lambda}\partial W_{\nu\lambda}/\partial y_\mu\ ,$$

or

$$[[y_\mu,\ y_\nu],\ y_\nu] + (W_{\mu\nu,\lambda} - \frac{1}{2}\,W_{\lambda\nu,\mu})\ \ [y_\lambda,\ y_\nu] = 0, \tag{37}$$

from which follows g^2 = const.

Although Eq. (33) is nonlinear, a class of solutions can be obtained with the ansatz

$$y_\mu = u_\mu\tau + f_\mu(\tau,\sigma),\quad u^2 = -1, u\cdot f = 0\ . \tag{38}$$

In particular, we can choose $u_\mu=(0,0,0,i)$, $f_\mu=f(\sigma)\times(\cos\tau,\sin\tau,0,0)$ and obtain

$$f'' - f(ff')' = 0\ , \tag{39}$$

which can be easily integrated. The solution represents a spinning rod. It differs from the corresponding solution in the Virasoro gauge by a redefinition of the coordinate σ.

An interesting possibility that suggests itself is to take the Poisson bracket notation in Eq. (35) seriously, and go to its "quantum mechanics" version, by regarding the internal coordinates τ and σ as noncommuting operators. It is totally unclear what this means, but we try it nevertheless. Some of the properties of the classical theory are lost. For example, g^2 will not automatically be a c-number, i.e., a constant. The difference stems from the fact that y_μ is in general a nonlinear function of τ and σ. Thus two exponentials $\exp[im\tau]$ and $\exp[in\sigma]$ commute if $mn/2\pi$ is an integer, but their classical Poisson bracket does not vanish.

An ansatz similar to Eq. (38) will be

$$f_x + if_y = e^{i\tau} f(\sigma), \quad f_x - if_y = f(\sigma)e^{-i\tau} . \tag{40}$$

We then get, in place of Eq. (39),

$$f''(\sigma) - \frac{1}{2} f(\sigma)[f(\sigma+1)^2 + f(\sigma-1)^2 - 2f(\sigma)^2] = 0. \tag{41}$$

An explicit solution of this equation has not been found yet.

We remark also that under the quantum mechanical interpretation, the ordinary action integral is replaced by

$$\mathrm{Tr}[y_\mu, y_\nu][y_\mu, y_\nu] . \tag{42}$$

By taking a variation δy_μ, we recover Eq. (35).

The structure of Eq. (35) reminds us of Maxwell's equations. In fact we can write them in an identical form by defining

$$Y_\mu \equiv p_\mu - e A_\mu(x) \equiv - i D_\mu ,$$

$$[x_\mu, p_\nu] = i . \tag{43}$$

Then

$$[Y_\mu, Y_\nu]/ei = - F_{\mu\nu} ,$$

$$[Y_\mu,[Y_\mu, Y_\nu]] = e \partial_\mu F_{\mu\nu} = 0 . \tag{44}$$

Observe the similarity between Eqs. (38) and (43). $A_\mu(x)$ is the analog of $f_\mu(\tau,\sigma)$. Since A_μ depends only on x, there is no difference between classical and quantum interpretations of $[x_\mu,p_\nu]$. The Maxwell Lagrangian in our language becomes

$$L = \frac{1}{4e^2} [Y_\mu, Y_\nu][Y_\mu, Y_\nu] . \tag{45}$$

Encouraged by this observation, we now try to include the sources. For this purpose, we will let $\hat{e}$ be a charge operator such that

$$[\hat{e},\psi] = -e \psi , \tag{46}$$

where ψ_i is a field carrying charge e. Then define

$$\hat{Y}_\mu \equiv p_\mu + \hat{e}\, A_\mu(x) \ , \tag{47}$$

so that

$$[\hat{Y}_\mu, \psi] = (-i\partial_\mu - e\, A_\mu)\psi . \tag{48}$$

If ψ is, for example, a scalar field, we add a Lagrangian

$$L_\psi = -[\hat{Y}_\mu, \psi]^\dagger [\hat{Y}_\mu, \psi] \ , \tag{49}$$

to L, Eq. (45). If $\hat{Y}$ instead of Y is used in the latter, however, we have to divide by $\hat{e}^2$ instead of e^2, or else project $\hat{e}$ onto a unit charge sector.

The non-Abelian generalization is straightforward. We write

$$\hat{Y}_\mu = p_\mu + \hat{e}^i\, A^i_\mu(x),$$

with

$$[\hat{e}^i, \hat{e}^j] = ig\, f^{ijk}\, \hat{e}^k . \tag{50}$$

Then

$$[\hat{Y}_\mu, \hat{Y}_\nu] = i\hat{e}^i\, F^i_{\mu\nu} \ , \tag{51}$$

$$L = \frac{1}{4Cg^2} [\hat{Y}_\mu, \hat{Y}_\nu][\hat{Y}_\mu, \hat{Y}_\nu]P \ . \tag{52}$$

The $\hat{e}^i$'s are g times the generators of a group. In Eq. (52), P stands for projection onto the space of an appropriate (e.g., fundamental) representation D of $\hat{e}^i$ such that $\mathrm{Tr}_D\, \hat{e}_i \hat{e}_k = \delta_{ik} C$.

It remains to be seen how one can exploit the observations made here, but there arise a few natural questions:

1. The parallelism between string and gauge field equations suggests that perhaps there is more than a formal connection between them. For example, the string may be a special realization of gauge fields in which some dynamical degrees of freedom are frozen while the others have become classical in a sense.

2. A non-Abelian string may be possible taking a cue from non-Abelian gauge fields.

3. Similarly, sources for a string may be introduced by following the way in which sources were introduced to gauge fields.

REFERENCES

1. M. Kalb and P. Ramond, Phys. Rev. D9, 2273 (1974); see also Y. Nambu, Proc. Symposium on Extended Systems in Field Theory, Paris, 1975 (Univ. of Chicago preprint EFI 75-43).

2. H. B. Nielsen and P. Olesen, Nucl. Phys. B61, 45 (1973).

3. E. Kyriakopoulos, Phys. Rev. 183, 1318 (1969).

4. Y. Nambu, Phys. Rev. D10, 4262 (1974) and Ref. 1.

5. See Y. Nambu, Lecture Notes for the Erice Workshops on Quark Models, 1975 (Univ. of Chicago preprint EFI 76-42).

6. A. Schild, Univ. of Texas preprint, 1976.

ADDENDUM

A more precise meaning of the charge operator $\hat{e}$ of Eq. (46) is as follows. It may be regarded as a 2×2 matrix which acts on the real and imaginary parts of $\psi(x)$. Then $\hat{e}^2 = e^2 \times 1$ (unit matrix). In rewriting Eq. (45) using $\hat{e}$, a trace average of the matrix should be taken. In the non-Abelian case, the $\hat{e}^i$ will be a set of matrices ($n \times n$ for SU(n)) acting, let us say, on a quark multiplet.

Collective Phenomena and Renormalization of Non-Linear Spinor Theories

TOHRU EGUCHI*†
The Enrico Fermi Institute
The University of Chicago
Chicago, Illinois

ABSTRACT

A new approach to collective phenomena and renormalization of non-linear spinor models is presented.

In this approach non-linear spinor models are converted into equivalent theories involving fermions and collective bosons by a simple rearrangement of perturbation series. For a wide class of non-linear spinor theories interactions among fermions and collective states are of the renormalizable kind and hence these models themselves turn out to be renormalizable.

The equivalence of various four-fermion theories to known renormalizable models is pointed out. The Nambu-Jona-Lasinio model, for instance, is shown to be equivalent to the linear σ-model.

Discussions are given on the origin of dynamical symmetry breakdown in spinor theories.

Non-linear spinor theories first proposed by Heisenberg,[1] Nambu-Jona-Lasinio[2] have been a useful theoretical laboratory to study collective phenomena in quantum field theories. As is well known these "superconductivity" models exhibit interesting phenomena, the emergence of collective bosonic states and the associated breakdown of symmetries. Here the agent of spontaneous symmetry breaking comes about as a bound state of the fundamental field of the theory and this is the mechanism which we need in the construction of a unified theory of weak and electromagnetic interactions.

---

*Enrico Fermi Fellow.

†Work supported in part by the NSF, Contract No. PHYS74-08833, and the Louis Block fund, the University of Chicago.

In spite of such theoretical appeal, however, the study of the superconductivity model has been confined to its crudest approximation, the Hartree-Fock approximation, in the past. This was due to the apparent non-renormalizable nature of the four-fermion interaction which made it difficult to handle higher order processes in a physically meaningful manner.

Today I would like to present a new approach to collective phenomena and renormalization of non-linear spinor models.[3] In this approach spinor theories are converted into theories of fermions and collective bosons by a simple rearrangement of perturbation series. Then it will be shown that for a wide class of non-linear spinor theories interactions among fermions and collective bosons are of the renormalizable kind and hence these models themselves are renormalizable by a well-defined prescription.

One example of such phenomena has been known for some time. A non-linear spinor theory of a vector-type self-interaction was studied by Bjorken[4] and others[5] some time ago and it was concluded that if a collective state is excited in the vector channel it behaves like a gauge field and the resulting theory is equivalent to spinor electrodynamics.

In this talk this kind of equivalence is extended to other theories. I will show that the Nambu-Jona-Lasinio model is equivalent to the linear σ-model, the model of ref. 6 to a broken U(1)×U(1) gauge theory and so on. In these examples S-matrices of the corresponding theories become identical, although their Green's functions are not necessarily the same because of certain differences in the definition of renormalization parameters.

In the following I will first quickly review the standard result of the Hartree-Fock treatment and then analyze its ultraviolet divergence structures. Next I introduce a path integral formulation and demonstrate the equivalence of four-fermion and Yukawa theories. Finally I make some discussions on the nature of the dynamical breakdown of symmetries.

I. HARTREE-FOCK RESULTS

Let us take the Nambu-Jona-Lasinio model as an example. The Lagrangian is given by

$$L = \bar{\psi}i\gamma\partial\psi + \frac{G_0}{2}\left((\bar{\psi}\psi)^2 + (\bar{\psi}i\gamma_5\vec{\tau}\psi)^2\right), \qquad (1\text{-}1)$$

where ψ is an iso-doublet spinor field. The theory has a chiral U(2)×SU(2) symmetry. The standard Hartree-Fock procedure is to add and subtract a fermion mass term to the Lagrangian,

$$L = \bar{\psi}i\gamma\partial\psi - m_0\bar{\psi}\psi + \frac{G_0}{2}((\bar{\psi}\psi)^2 + (\bar{\psi}i\gamma_5\vec{\tau}\psi)^2) + m_0\bar{\psi}\psi, \quad (1\text{-}2)$$

and to regard the first two terms as the free part of the Lagrangian. Then the self-energy effects caused by the last two terms are assumed to cancel,

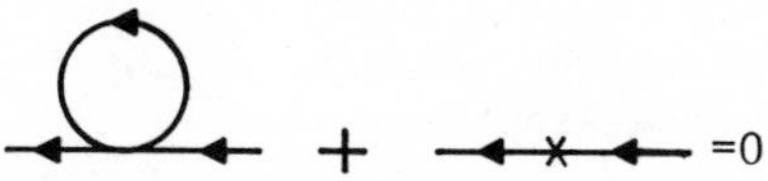

$$=0 \quad (1\text{-}3)$$

and one obtains the self-consistency equation,

$$\frac{2G_0 i}{m_0}\int \frac{d^4p}{(2\pi)^4}\frac{\mathrm{Tr}1}{\not{p}-m_0} = 8G_0 i\int \frac{d^4p}{(2\pi)^4}\frac{1}{p^2-m_0^2} = 1. \quad (1\text{-}4)$$

This equation can be interpreted as determining the "vacuum expectation value" m_0 in terms of G_0 and the ultraviolet cut-off Λ.

Now it is easy to observe the existence of various bosonic collective states. In the scalar channel, we have,

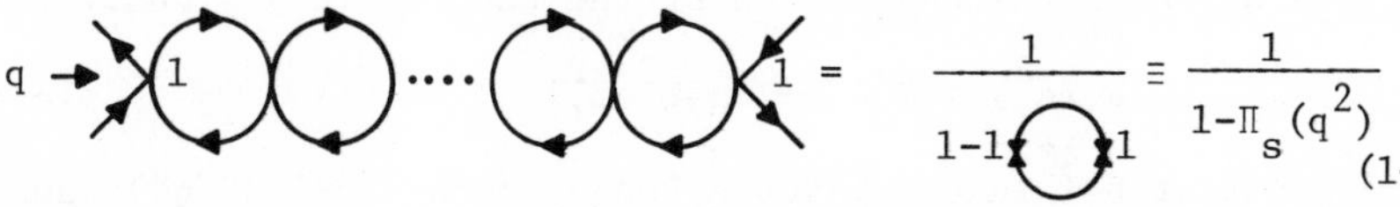

$$q \rightarrow 1 \cdots 1 = \frac{1}{1-1\,\bigcirc\,1} \equiv \frac{1}{1-\Pi_s(q^2)}, \quad (1\text{-}5)$$

where

$$\Pi_s(q^2) = 2G_0 i\int \frac{d^4p}{(2\pi)^4}\,\mathrm{Tr}\,\frac{1}{\not{p}-m_0}\,\frac{1}{\not{p}-\not{q}-m_0}. \quad (1\text{-}6)$$

Because of the self-consistency equation (1-4),

$$\Pi_s(4m_0^2) = 1, \quad (1\text{-}7)$$

hence

$$\Pi_s(q^2) = 1 + C_s(q^2-4m_0^2) + \cdots\cdots. \quad (1\text{-}8)$$

Similarly in the pseudo-scalar channel we have,

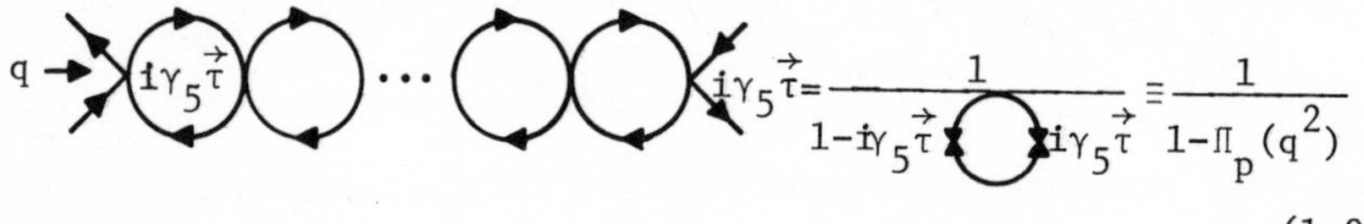

$$q \rightarrow i\gamma_5\vec{\tau} \cdots i\gamma_5\vec{\tau} = \frac{1}{1-i\gamma_5\vec{\tau}\,\bigcirc\,i\gamma_5\vec{\tau}} \equiv \frac{1}{1-\Pi_p(q^2)}, \quad (1\text{-}9)$$

$$\Pi_p(q^2) = 2G_0 i \int \frac{d^4p}{(2\pi)^4} \mathrm{Tr}\, i\gamma_5 \frac{1}{\not{p}-m_0} i\gamma_5 \frac{1}{\not{p}-\not{q}-m_0} = 1 + C_p q^2 + \cdots , \tag{1-10}$$

$$\Pi_p(0) = 1. \tag{1-11}$$

In both cases the infinite sum of iterated bubbles gives rise to a boson propagator.

$$\cdots (i \bigcirc\!\bigcirc \cdots \bigcirc\!\bigcirc i) \cdots \equiv \sim\!\sim\!\sim = \frac{1}{1-\Pi_i(q^2)} = \frac{1}{-C_i(q^2-m_i^2)+\cdots} . \tag{1-12}$$

So far I have been cavalier about ultraviolet divergences. If the momentum integration in Eq. (1-4) is performed, one obtains,

$$G_0\Lambda^2 \sim 1 . \tag{1-13}$$

On the other hand the evaluation of the bubble $\Pi(q^2)$ gives,

$$\Pi_i(q^2) \sim G_0\Lambda^2 + G_0(q^2-m_i^2)\ln\Lambda + \cdots . \tag{1-14}$$

Hence the factor 1 in the inverse boson propagator $1-\Pi(q^2)$ can be interpreted as a mass renormalization counter term,

$$1-\Pi(q^2) = G_0\Lambda^2 - (G_0\Lambda^2 + G_0(q^2-m_i^2)\ln\Lambda + \cdots) = -G_0\ln\Lambda(q^2-m_i^2). \tag{1-15}$$

Thus the self-consistency Eq. (1-4) serves also as the boson mass renormalization condition. Elimination of the remaining divergence, $\ln\Lambda$ in the boson wave-function renormalization, is not performed in the Hartree-Fock approximation.

Here we note, however, a strong cancellation of ultraviolet divergences is already taking place at this stage of the theory. In fact if we evaluate term by term in the series of repeated bubbles, we obtain badly divergent quantities

$$\sum_{n=1}^{\infty} \overbrace{\bigcirc\!\bigcirc \cdots \bigcirc\!\bigcirc}^{n} = \frac{1}{1-\bigcirc} = \frac{1}{\ln\Lambda} . \tag{1-16}$$

$$(\Lambda^2)^n \qquad\qquad \Lambda^2 \quad \Lambda^2$$

When they are summed, however, we are left only with $\ln\Lambda$.

II. DIVERGENCE STRUCTURE OF THE THEORY

It is well known that a four-fermion theory has a badly divergent ultraviolet structure. It is easy, however, to isolate the origin of severe divergences because of the simplicity of the topology of four-fermion diagrams. Indeed all the divergences of the theory worse than those in the renormalizable models are created by the repetition of the basic bubble diagram.

On the other hand, as we have seen above, ultraviolet behavior of iterated bubbles becomes very much improved if a summation is performed over their infinite series. Hence it seems possible to significantly reduce the ultraviolet divergences of a four-fermion theory by first making a rearrangement of its perturbation series and then summing over its repeated bubbles. Let us next see that this is indeed the case in various types of diagrams.

First let us examine the fermion self-energy type graphs,

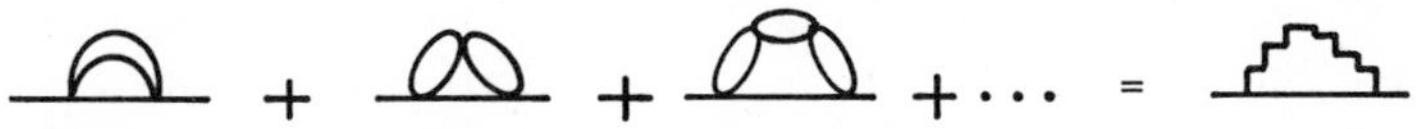

(2-1)

In fact after the summation all the divergences worse than $\ell n\Lambda$ disappear. The additional loop integration gives the familiar $\ell n\Lambda$ of the fermion mass renormalization.

A similar situation obtains in the case of boson self-energy and vertex correction graphs,

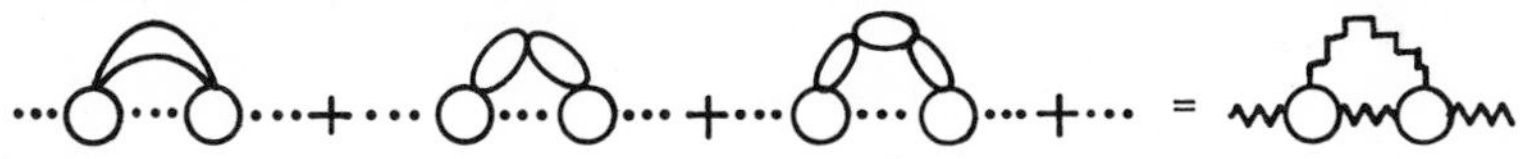

(2-2)

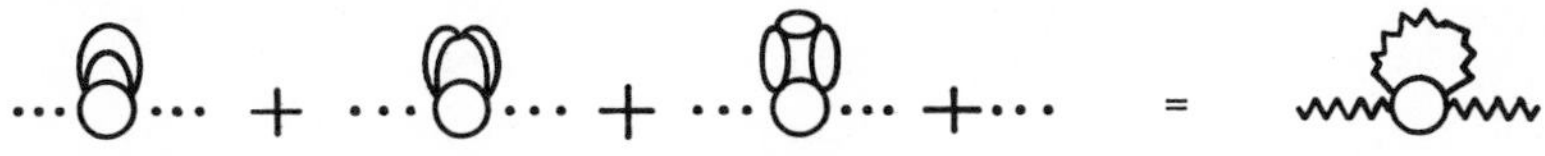

(2-3)

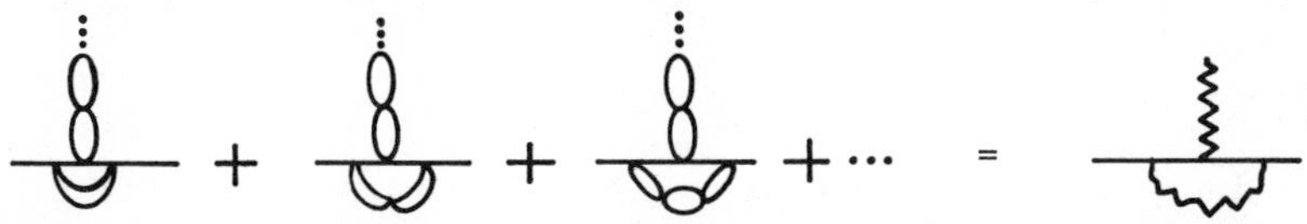

(2-4)

Again after the summation severe divergences cancel and the resulting graphs are the familiar ones in Yukawa theories.

We may consider even more complicated configurations. For instance, diagrams

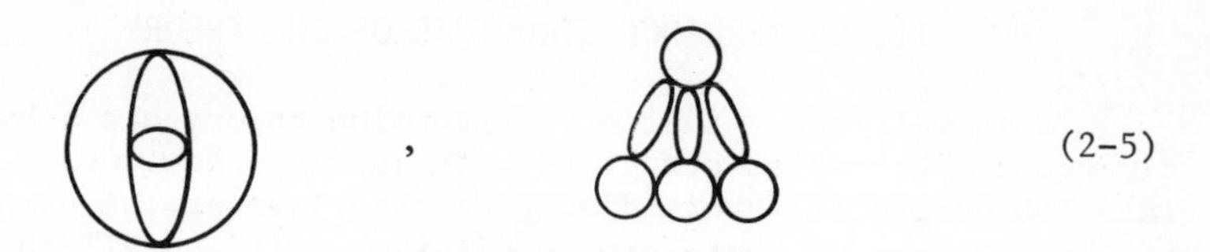

are included in well-behaved graphs,

As we see in these examples, after rearrangement and summation the topology and convergence properties of diagrams become identical to those of the renormalizable Yukawa theories. No worse divergences remain. Therefore we expect that a simple modification of the perturbation series will cast a four-fermion theory into a manifestly renormalizable form. As we see later this is most conveniently done by a path integral technique.

Before going into discussions using a functional method, let us next consider a particular treatment of a four-fermion theory which clarifies the nature of the Hartree-Fock approximation. We analyze the following limit of a four-fermion theory where,

$$G_0\Lambda^2 = \text{fixed},$$

and

$$\Lambda \to \infty \,. \qquad (2\text{-}7)$$

Apparently in this treatment a rearrangement of perturbation series is implied since Λ is counted as $G_0^{-\frac{1}{2}}$. The basic bubble graph is regarded as $(G_0)^0$ and hence by its repetition we have collective boson propagators already at the "unperturbed" level of the theory.

On the other hand all the other diagrams vanish in this limit. This is because those graphs contain at least one $\ell n\Lambda$ (or Λ^0) in place of Λ^2 and hence $\ell n\Lambda/\Lambda^2 \to 0$ (or $1/\Lambda^2 \to 0$) as $\Lambda \to \infty$. This was actually the case of radiative correction graphs Eqs. (2-1)-(2-4). Similarly boson scattering diagrams are vanishing because of the loop integration at the center,

$$(2\text{-}8)$$

Hence the theory has a tree structure in this limit and is essentially free except for the single boson exchange between fermions.

The situation can be schematically summarized as follows.

$$\text{fermion} \xrightarrow[\text{pairing force}]{\Lambda^2} \text{boson}$$

$$\text{boson} \xrightarrow[\ell n\Lambda]{\text{radiative reaction}} \text{fermion} \qquad (2\text{-}9)$$

Since in the above limit Eq. (2-7) only the pairing interactions among fermions are taken into account while the radiative reactions of bosons are neglected, this is nothing but the conventional Hartree-Fock approximation. Thus the Hartree-Fock method is a tree approximation to the "true" theory. If we go beyond the Hartree-Fock treatment, we will observe boson-boson scatterings as well as all kinds of radiative corrections to the tree approximation.

III. PATH INTEGRAL METHOD

Now we are ready to introduce a path integral formulation. In the path integral method the generating functional of the theory is given by

$$W[\eta,\bar{\eta}] = \frac{1}{N}\iint \exp\, i(L(\psi,\bar{\psi}) + \bar{\eta}\psi + \bar{\psi}\eta)d\psi d\bar{\psi}\ . \qquad (3\text{-}1)$$

N is a normalization factor which will be suppressed hereafter. Next we introduce a new Lagrangian L' and integration variables σ and $\vec{\pi}$ in such a way that the above generating functional is expressed as

$$W[\eta,\bar{\eta}] = \int \exp\, i(L'(\psi,\bar{\psi},\sigma,\vec{\pi}) + \bar{\eta}\psi + \bar{\psi}\eta)d\psi d\bar{\psi}d\sigma d\vec{\pi}. \qquad (3\text{-}2)$$

L' is easily found and given by

$$L' = \bar{\psi}(i\gamma\partial - g_0(\sigma + i\gamma_5\vec{\pi}\vec{\tau})\psi - \frac{1}{2}\,\delta\mu_0^2(\sigma^2 + \vec{\pi}^2). \tag{3-3}$$

Here g_0 and $\delta\mu_0$ are related to G_0 by

$$G_0 = g_0^2/\delta\mu_0^2. \tag{3-4}$$

From now on let us compare our theory with the linear σ-model whose Lagrangian is given by

$$L_\sigma = \bar{\psi}(i\gamma\partial - g_0'(\sigma + i\gamma_5\vec{\pi}\vec{\tau}))\psi + \frac{1}{2}((\partial_\mu\sigma)^2 + (\partial_\mu\vec{\pi})^2 - \mu'^2(\sigma^2+\vec{\pi}^2))$$
$$- \frac{\lambda_0'}{4}(\sigma^2+\vec{\pi}^2)^2 - \frac{1}{2}\,\delta\mu_0'^2(\sigma^2+\vec{\pi}^2)\ . \tag{3-5}$$

Although in our Lagrangian Eq. (3-3) boson kinetic and interaction terms in L_σ are missing, they are actually present in the four-fermion theory as we know from our previous diagrammatic analysis. I will show in the following that these missing terms are in fact created out of radiative corrections in the path integral formulation and S-matrices of the two theories L' and L_σ become identical to each other.

In order to demonstrate this let us next perform integrations over fields ψ and $\bar{\psi}$ in Eq. (3-2). Using the standard formula we obtain,

$$W[\eta,\bar{\eta}] = \iint d\sigma d\vec{\pi}\ \exp\ i(-\frac{1}{2}\,\delta\mu_0^2(\sigma^2+\vec{\pi}^2)$$
$$- i\mathrm{Tr}\,\ell n[1 - \frac{1}{i\gamma\partial}\,g_0(\sigma+i\gamma_5\vec{\pi}\vec{\tau})]$$
$$+ \bar{\eta}\,\frac{1}{i\gamma\partial - g_0(\sigma+i\gamma_5\vec{\pi}\vec{\tau})}\,\eta). \tag{3-6}$$

The second term is expanded into a power series in g_0,

$$U' \equiv -i\mathrm{Tr}\,\ell n[1 - \frac{1}{i\gamma\partial}\,g_0(\sigma+i\gamma_5\vec{\pi}\vec{\tau})] = \sum_{n=1}^{\infty} U'^{(n)}, \tag{3-7}$$

where

$$U'^{(n)} \equiv \frac{i}{n}\,\mathrm{Tr}(\frac{1}{i\gamma\partial}\,g_0(\sigma+i\gamma_5\vec{\pi}\vec{\tau}))^n. \tag{3-8}$$

In this expansion, however, we have a trouble of infrared divergences coming from the masslessness of the spinor field. Hence we introduce a shifted field s by

$$\sigma = s + v_0, \tag{3-9}$$

and treat g_0v_0 as a bare fermion mass. The vacuum value v_0 is determined by the requirement that there should be no linear term in s in our effective action. Now we have the expansion,

$$W[\eta,\bar{\eta}]=\iint d\sigma d\vec{\pi}\ \exp\ i(-\frac{1}{2}\ \delta\mu_0^2(\sigma^2+\vec{\pi}^2)+U+\bar{\eta}\ \frac{1}{i\gamma\partial-g_0(\sigma+i\gamma_5\vec{\pi}\vec{\tau})}\ \eta), \tag{3-10}$$

$$U \equiv -i\mathrm{Tr}\ell n\ [1-\frac{1}{i\gamma\partial-g_0v_0}\ g_0(s+i\gamma_5\vec{\pi}\vec{\tau})] = \sum_{n=1}^{\infty} U^{(n)} \tag{3-11}$$

$$U^{(n)} \equiv \frac{i}{n}\ \mathrm{Tr}\ (\frac{1}{i\gamma\partial-g_0v_0}\ g_0(s+i\gamma_5\vec{\pi}\vec{\tau}))^n. \tag{3-12}$$

A diagrammatic expression of this series is given by

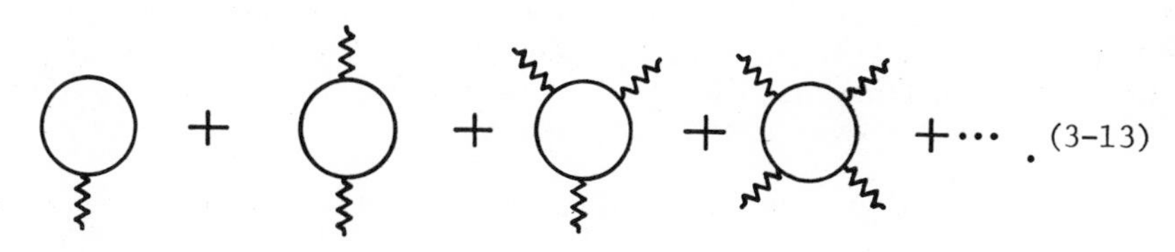
(3-13)

We note that these diagrams happen to be the same as those of the lowest-order radiative corrections to the effective action in the conventional formulation of the functional method. In our treatment, however, the external lines of the above diagrams are quantized fields s and $\vec{\pi}$ and not their expectation values. In the present case $U^{(i)}$ (i=1,2,3,4) give ultraviolet divergences while $U^{(n)}$ ($n\geqslant5$) are all convergent. After an explicit evaluation of the divergent part of the diagrams, the exponent of Eq. (3-10) is given by

$$\begin{aligned}\bar{L} = &-\frac{1}{2}(\delta\mu_0^2-2I_2g_0^2)(\sigma^2+\vec{\pi}^2) + I_0g_0^2\ \frac{1}{2}((\partial_\mu\sigma)^2+(\partial_\mu\vec{\pi})^2)\\ &+ I_0g_0^2(g_0v_0)^2(\sigma^2+\vec{\pi}^2) - \frac{1}{2}\ I_0g_0^4(\sigma^2+\vec{\pi}^2)^2\\ &+ L_c(g_0,v_0) + \bar{\eta}\ \frac{1}{i\gamma\partial-g_0(\sigma+i\gamma_5\vec{\pi}\vec{\tau})}\ \eta,\end{aligned} \tag{3-14}$$

where

$$L_c = \sum_{i=2}^{4} U_c^{(i)} + \sum_{i=5}^{\infty} U^{(i)}, \tag{3-15}$$

and $U_c^{(i)}$ is the convergent part of $U^{(i)}$. Separation of $U_c^{(i)}$ (i=2,3,4) into convergent and divergent parts is done in a convenient manner. I_2 and I_0 are quadratically and logarithmically divergent integrals, respectively,

$$I_2 = 4i \int \frac{d^4p}{(2\pi)^4} \frac{1}{p^2 - (g_0 v_0)^2} , \qquad (3\text{-}16)$$

$$I_0 = -4i \int \frac{d^4p}{(2\pi)^4} \frac{1}{(p^2 - (g_0 v_0)^2)^2} . \qquad (3\text{-}17)$$

In deriving Eq. (3-14) we have first calculated diagrams using the field s and in the end we have eliminated s in favor of σ. Separation of self-energy terms into two is done in such a way that in the sum of third and fourth terms of $\bar{L}$ no linear term in s exists. Although Eq. (3-14) contains divergent coefficients we have in fact created kinetic and interaction terms of bosons out of radiative corrections. In order to find out an appropriate renormalization prescription for Eq. (3-14) let us next perform a similar calculation in the case of the σ-model. After the integration over ψ and $\bar{\psi}$ fields in

$$W_\sigma[\eta,\bar{\eta}] = \int \exp\, i(L_\sigma(\psi,\bar{\psi},\sigma,\vec{\pi}) + \bar{\eta}\psi + \bar{\psi}\eta)d\psi d\bar{\psi} d\sigma d\vec{\pi} \qquad (3\text{-}18)$$

we find that the exponent is given by

$$\begin{aligned} \bar{L}_\sigma = & -\frac{1}{2}(\delta\mu_0'^2 + \mu'^2 + \lambda_0' v_0'^2 - 2I_2' g_0'^2)(\sigma^2 + \vec{\pi}^2) \\ & + (1 + I_0' g_0'^2)\frac{1}{2}((\partial_\mu\sigma)^2 + (\partial_\mu\vec{\pi})^2) \\ & + \frac{1}{2}(\lambda_0' v_0'^2 + 2I_0' g_0'^2 (g_0' v_0')^2)(\sigma^2 + \vec{\pi}^2) \\ & - \frac{1}{4}(\lambda_0' + 2I_0' g_0'^4)(\sigma^2 + \vec{\pi}^2)^2 + L_c(g_0', v_0') \\ & + \bar{\eta} \frac{1}{i\gamma\partial - g_0'(\sigma + i\gamma_5 \vec{\pi}\vec{\tau})} \eta . \end{aligned} \qquad (3\text{-}19)$$

Here $L_c(g_0', v_0')$ is obtained from $L_c(g_0, v_0)$ by substituting g_0' and v_0' for g_0 and v_0. Similarly I_2' and I_0' are given by Eqs. (3-16) and (3-17) with g_0 and v_0 replaced by g_0' and v_0'. Note that $\bar{L}_\sigma$ has exactly the same operatorial structure as $\bar{L}$.

Renormalization prescriptions for the linear σ-model are well-known. By the introduction of wave function and vertex renormalization factors Z_M and Z_λ, infinities in Eq. (3-19) are absorbed into renormalization parameters,

$$1 + I_0' g_0'^2 = \frac{1}{Z_M} , \tag{3-20}$$

$$\lambda_0' + 2I_0' g_0'^4 = \frac{\lambda_0}{Z_\lambda} \tag{3-21}$$

$$-\delta\mu_0'^2 - \mu'^2 - \lambda_0' v_0'^2 + 2I_2' g_0'^2 = 0. \tag{3-22}$$

Defining renormalized fields and vertices as

$$\sigma_R = Z_M^{-1/2} \sigma , \quad \vec{\pi}_R = Z_M^{-1/2} \vec{\pi} , \tag{3-23}$$

$$v' = Z_M^{-1/2} v_0' , \tag{3-24}$$

$$\lambda' = Z_\lambda^{-1} Z_M^2 \lambda_0' , \tag{3-25}$$

we obtain the following expression of $\bar{L}_\sigma$,

$$\bar{L}_\sigma = \frac{1}{2} ((\partial_\mu \sigma_R)^2 + (\partial_\mu \vec{\pi}_R)^2 + \lambda' v'^2 (\sigma_R^2 + \vec{\pi}_R^2))$$
$$- \frac{\lambda'}{4} (\sigma_R^2 + \vec{\pi}_R^2)^2 + L_c(g_0', v_0')$$
$$+ \bar{\eta} \frac{1}{i\gamma\partial - g_0'(\sigma + i\gamma_5 \vec{\pi}\vec{\tau})} \eta . \tag{3-26}$$

Referring to the above procedure, we find that our theory should be renormalized as follows,

$$I_0 g_0^2 = \frac{1}{Z_M} , \tag{3-27}$$

$$2I_0 g_0^4 = \frac{\lambda_0}{Z_\lambda} , \tag{3-28}$$

$$-\delta\mu_0^2 + 2I_2 g_0^2 = 0 , \tag{3-29}$$

$$\sigma_R = Z_M^{-1/2} \sigma, \quad \vec{\pi}_R = Z_M^{-1/2} \vec{\pi} , \tag{3-30}$$

$$v = Z_M^{-1/2} v_0 , \tag{3-31}$$

$$\lambda = Z_\lambda^{-1} Z_M^2 \lambda_0 . \tag{3-32}$$

By Eq. (3-28) we have introduced a new parameter λ_0 into our theory. Note that Eq. (3-29) is exactly the same as the self-consistency Eq. (1-4). Then the exponent $\bar{L}$ is given by

$$\bar{L} = \frac{1}{2}\left((\partial_\mu \sigma_R)^2 + (\partial_\mu \vec{\pi}_R)^2 + \lambda v^2(\sigma_R^2 + \vec{\pi}_R^2)\right) - \frac{\lambda}{4}(\sigma_R^2 + \vec{\pi}_R^2)^2 + L_c(g_0, v_0) + \bar{\eta}\,\frac{1}{i\gamma\partial - g_0(\sigma + i\gamma_5\vec{\pi}\vec{\tau})}\,\eta \ . \tag{3-33}$$

Hence generating functionals of two theories have exactly the same structure,

$$W = \mathcal{W}(g_0, \lambda_0, v_0) \ , \tag{3-34}$$

$$W_\sigma = \mathcal{W}(g_0', \lambda_0', v_0') \tag{3-35}$$

where $\mathcal{W}$ is a certain functional. Differences between Eqs. (3-20)∿(3-22) and Eqs. (3-27)∿(3-29) are easy to understand. Since originally we had no kinetic terms nor interaction terms for bosons in our theory, this explains the absence of terms like 1 or λ_0 in the left-hand sides of Eqs. (3-27)∿(3-29). These missing terms, however, have now been created out of radiative corrections and the difference between the two models has been absorbed into the relations between renormalization parameters and bare quantities. Since no renormalization parameters nor bare quantities appear in S-matrix elements, both theories predict the same S-matrix to all orders in perturbation theory.

A convenient feature of the above method lies in the fact that only those calculations corresponding to lowest-order radiative corrections are needed in order to demonstrate the equality of the S-matrix to all orders in perturbation theory. There is, however, a more formal argument by means of which we can infer the equivalence even without doing any explicit calculations. In fact as we see from the foregoing discussions our essential observation is to exploit the ambiguities in decomposing renormalized quantities into bare and counter terms. We can rewrite, for instance, the renormalized Lagrangian of the σ-model in terms of bare quantities in two different ways,

$$L_{\sigma R}=\bar{\psi}_R i\gamma\partial\psi_R-g\bar{\psi}_R(\sigma_R+i\gamma_5\vec{\pi}_R\vec{\tau})\psi_R+\frac{1}{2}((\partial_\mu\sigma_R)^2+(\partial_\mu\vec{\pi}_R)^2-\mu^2(\sigma_R^2+\vec{\pi}_R^2))$$

$$-\frac{\lambda}{4}(\sigma_R^2+\vec{\pi}_R^2)^2 , \tag{3-36}$$

$$=[\bar{\psi}i\gamma\partial\psi-g_0\bar{\psi}(\sigma+i\gamma_5\vec{\pi}\vec{\tau})\psi+\frac{1}{2}((\partial_\mu\sigma)^2+(\partial_\mu\vec{\pi})^2-\mu^2(\sigma^2+\vec{\pi}^2))$$

$$-\frac{1}{2}\delta\mu_0^2(\sigma^2+\vec{\pi}^2)-\frac{\lambda_0}{4}(\sigma^2+\vec{\pi}^2)^2]$$

$$+[(\frac{1}{Z_F}-1)\bar{\psi}i\gamma\partial\psi-(\frac{1}{Z_g}-1)g_0\bar{\psi}(\sigma+i\gamma_5\vec{\pi}\vec{\tau})\psi$$

$$+(\frac{1}{Z_M}-1)\frac{1}{2}((\partial_\mu\sigma)^2+(\partial_\mu\vec{\pi})^2-\mu^2(\sigma^2+\vec{\pi}^2))+\frac{1}{2}\delta\mu_0^2(\sigma^2+\vec{\pi}^2)$$

$$-\frac{\lambda_0}{4}(\frac{1}{Z_\lambda}-1)(\sigma^2+\vec{\pi}^2)^2], \tag{3-37}$$

$$=[\bar{\psi}i\gamma\partial\psi-g_0\bar{\psi}(\sigma+i\gamma_5\vec{\pi}\vec{\tau})\psi-\frac{1}{2}\delta\mu_0^2(\sigma^2+\vec{\pi}^2)]+[(\frac{1}{Z_F}-1)\bar{\psi}i\gamma\partial\psi$$

$$-(\frac{1}{Z_g}-1)g_0\bar{\psi}(\sigma+i\gamma_5\vec{\pi}\vec{\tau})\psi+\frac{1}{Z_M}\frac{1}{2}((\partial_\mu\sigma)^2+(\partial_\mu\vec{\pi})^2-\mu^2(\sigma^2+\vec{\pi}^2))$$

$$+\frac{1}{2}\delta\mu_0^2(\sigma^2+\vec{\pi}^2)-\frac{\lambda_0}{4}\frac{1}{Z_\lambda}(\sigma^2+\vec{\pi}^2)^2] . \tag{3-38}$$

Equation (3-37) is the usual prescription where terms in the first bracket are regarded as the bare Lagrangian of the theory and those in the second one as its radiative corrections. On the other hand Eq. (3-38) corresponds to our treatment where the bare Lagrangian consists only of the three terms of the first bracket. Of course such kind of ambiguities in quantum field theories is well-known and forms the basis of the renormalization group approach. Its relevance in non-linear spinor theories, however, has not been well appreciated. If we take a variation of the boson fields in the bare Lagrangian of Eq. (3-37) we obtain the usual equation of motion,

$$\Box\sigma+(\mu^2+\delta\mu_0^2)\sigma+g_0\bar{\psi}\psi+\lambda_0\sigma(\sigma^2+\vec{\pi}^2)=0, \tag{3-39}$$

and a similar one for $\vec{\pi}$. On the other hand the bare Lagrangian of Eq. (3-38) gives us the Euler equations,

$$\delta\mu_0^2\sigma+g_0\bar{\psi}\psi=0, \tag{3-40}$$

$$\delta\mu_0^2\vec{\pi} + g_0\bar{\psi}i\gamma_5\vec{\tau}\psi = 0 \ . \tag{3-41}$$

In Eq. (3-39) σ field has its own degrees of freedom while in Eqs. (3-40) and (3-41) boson fields appear to be entirely dependent on the spinor field. Thus boson fields are at the same time elementary and composite.

Apparently the above arguments can be applied to theories other than the σ-model and most of renormalizable models involving fermions and bosons are realized as various kinds of non-linear spinor theories.

IV. SPONTANEOUSLY BROKEN GAUGE THEORIES

Let us next consider consequences of possible symmetries of non-linear spinor theories. If the primary interaction of a four-fermion theory possesses a certain symmetry, its equivalent renormalizable theory will also exhibit a corresponding symmetry. Here the interesting phenomenon is that sometimes the latter induced symmetry turns out to be higher than the original one. In particular whenever a collective mode is excited in a channel of a conserved current we necessarily arrive at a local gauge symmetry starting from a globally invariant one. Let us next see how this happens in the model of Bjorken.[4,5]

The model is given by the following Lagrangian,

$$L = \bar{\psi}(i\gamma\partial - m_0)\psi - \frac{G_0'}{2}(\bar{\psi}\gamma_\mu\psi)(\bar{\psi}\gamma^\mu\psi), \tag{4-1}$$

having a global U(1) symmetry. Applying the same technique as in the previous section we obtain

$$L' = \bar{\psi}(i\gamma\partial - m_0)\psi - e_0\bar{\psi}\gamma_\mu\psi A^\mu + \frac{1}{2}\delta\mu_0^2 A_\mu A^\mu, \tag{4-2}$$

where

$$G_0' = \frac{e_0^2}{\delta\mu_0^2} \ . \tag{4-3}$$

The equivalence of this theory to QED is well-known. In fact, if we perform a similar calculation as in the σ-model, we find

$$\bar{L} = \frac{1}{2}\left(\delta\mu_0^2 - \frac{1}{2}e_0^2(I_2 + m_0^2 I_0)\right) + \left(\frac{1}{3}e_0^2 I_0\right)\left(-\frac{1}{4}F_{\mu\nu}F^{\mu\nu}\right) + L_c$$
$$+ \text{(fermion source term)} \ . \tag{4-4}$$

Here $F_{\mu\nu}$ is the usual curl of A_μ and we have explicitly exhibited the photon mass term coming from the lowest-order vacuum polarization diagram. This is to be canceled against the photon mass counter term 1/2 $\delta\mu_0^2 A_\mu A^\mu$. Then via the renormalization prescription,

$$\delta\mu_0^2 - \frac{1}{2} e_0^2(I_2+m_0^2 I_0) = 0, \qquad (4\text{-}5)$$

$$\frac{1}{3} e_0^2 I_0 = \frac{1}{Z_3}, \qquad (4\text{-}6)$$

ultraviolet infinities are eliminated and we obtain

$$\bar{L} = -\frac{1}{4} F_{\mu\nu R}F^{\mu\nu R} + L_c + \text{(fermion source term)}, \qquad (4\text{-}7)$$

which has exactly the same structure as the corresponding expression in QED.

From the above example it is already apparent how we can create a local gauge symmetry starting from a global invariance. In Eq. (4-2) we have a collective excitation A_μ which is coupled to a conserved current $\bar{\psi}\gamma_\mu\psi$. Hence L is invariant under

$$A_\mu \rightarrow A_\mu + \partial_\mu\Lambda, \qquad (4\text{-}8)$$

with an arbitrary Λ. Although the mass counter term 1/2 $\delta\mu_0^2 A_\mu A^\mu$ seems to spoil this invariance, it in fact kills the gauge non-invariant photon mass term coming from radiative corrections and saves the gauge invariance. (If in the above we had used a gauge invariant regulator, we would not have created a photon mass term. In this case we have to put $\delta\mu_0=0$. This choice is a bit awkward but is known to be consistent. Then we have a trivial equation 0=0 instead of Eq. (4-5).)

The non-Abelian analogue of the above mechanism is quite similar. If we take as our starting Lagrangian,

$$L = \bar{\psi}(i\gamma\partial - m_0)\psi - \frac{G_0'}{2}(\bar{\psi}\gamma_\mu\lambda_\alpha\psi)(\bar{\psi}\gamma^\mu\lambda_\alpha\psi), \qquad (4\text{-}9)$$

$$\lambda_\alpha \text{ ; SU(n) matrices,}$$

and introduce an excitation A_μ^α in the channel $\bar{\psi}\gamma_\mu\lambda_\alpha\psi$, then it behaves exactly like an SU(n) Yang-Mills field and we arrive at the so-called quantum chromodynamics.

With the above preliminaries it is now possible to tell which kind of non-linear spinor theories lead to which kind of renormalizable models even without going into detailed calculations. Let us take the model considered in ref. 6 for example. It is given by

$$L = \bar{\psi}i\gamma\partial\psi + \frac{G_0}{2}((\bar{\psi}\psi)^2 + (\bar{\psi}i\gamma_5\psi)^2)$$
$$- \frac{G_0'}{2}((\bar{\psi}\gamma_\mu\psi)(\bar{\psi}\gamma^\mu\psi)+(\bar{\psi}\gamma_5\gamma_\mu\psi)(\bar{\psi}\gamma_5\gamma^\mu\psi)). \quad (4\text{-}10)$$

The theory possesses a symmetry U(1) × U(1) and hence a conserved vector and axial-vector current. Therefore if we introduce collective fields in the following way,

$$L' = \bar{\psi}i\gamma\partial\psi - g_0\,\bar{\psi}(\phi_s+i\gamma_5\phi_p)\psi - e_0\,\bar{\psi}(\gamma_\mu V^\mu+\gamma_5\gamma_\mu A^\mu)\psi$$
$$+ \text{(mass counter terms)}, \quad (4\text{-}11)$$

then V_μ and A_μ behave like a vector and axial-vector gauge field. Since mesons ϕ_s, ϕ_p, V_μ and A_μ do not carry a baryonic charge, V_μ will couple only to the spinor field. On the other hand under a chiral rotation fields transform as,

$$\psi \to e^{i\alpha\gamma_5}\psi, \quad (4\text{-}12)$$

$$\phi \equiv \frac{\phi_s+i\phi_p}{\sqrt{2}} \to e^{-2i\alpha}\phi, \quad (4\text{-}13)$$

$$V_\mu,\ A_\mu \to V_\mu,\ A_\mu. \quad (4\text{-}14)$$

Hence both the spinor and ϕ field are axially charged and interact with A_μ. Since ϕ develops a vacuum expectation value like in the σ-model, A_μ acquires a mass due to the Higgs mechanism. Therefore our model will be equivalent to a broken gauge theory of chiral U(1) × U(1) symmetry where only the axial U(1) gauge is spontaneously broken. In fact after a detailed calculation we find that the equivalent theory is described by the Lagrangian

$$L_{\text{Higgs}} = -\frac{1}{4}F_{\mu\nu}F^{\mu\nu} - \frac{1}{4}G_{\mu\nu}G^{\mu\nu} + i\bar{\psi}\gamma\partial\psi + |\partial_\mu\phi+2ie_0A_\mu|^2$$
$$-g_0\bar{\psi}(\phi_s+i\gamma_5\phi_p)\psi-e_0\bar{\psi}(\gamma_\mu V^\mu+\gamma_5\gamma_\mu A^\mu)\psi-\mu_0^2|\phi|^2-\lambda_0|\phi|^4, \quad (4\text{-}15)$$

where

$$F_{\mu\nu} = \partial_\mu V_\nu - \partial_\nu V_\mu, \quad (4\text{-}16)$$

$$G_{\mu\nu} = \partial_\mu A_\nu - \partial_\nu A_\mu. \quad (4\text{-}17)$$

It is also easy to discuss the non-Abelian analogue of the above example.[7] The Lagrangian is given by

$$L = \bar{\psi}i\gamma\partial\psi + \frac{G_0}{2}\left((\bar{\psi}\psi)^2+(\bar{\psi}i\gamma_5\lambda_\alpha\psi)^2+(\bar{\psi}\lambda_\alpha\psi)^2+(\bar{\psi}i\gamma_5\psi)^2\right)$$
$$- \frac{G_0'}{2}\left((\bar{\psi}\gamma_\mu\psi)(\bar{\psi}\gamma^\mu\psi)+(\bar{\psi}\gamma_\mu\lambda_\alpha\psi)(\bar{\psi}\gamma^\mu\lambda_\alpha\psi)+(\bar{\psi}\gamma_5\gamma_\mu\psi)(\bar{\psi}\gamma_5\gamma^\mu\psi)\right.$$
$$\left.+(\bar{\psi}\gamma_5\gamma_\mu\lambda_\alpha\psi)(\bar{\psi}\gamma_5\gamma^\mu\lambda_\alpha\psi)\right). \qquad (4\text{-}18)$$

The theory has a symmetry $U(n) \times U(n)$. After a similar treatment as in the previous case, the model is shown to be equivalent to a spontaneously broken gauge theory of chiral $U(n) \times U(n)$ symmetry.

Physical spectra of the theory we obtain via the Higgs mechanism depend on the choice of vacuum expectation values. In the case of n=2, for instance, if we take $\langle\phi_S\rangle \neq 0$ and $\langle\vec{\phi}_S\rangle = 0$ then we obtain a theory of massive ϕ_S, $\vec{\phi}_S$, A_μ and $\vec{A}_\mu$ fields interacting with massless V_μ and $\vec{V}_\mu$ mesons. In this choice the axial U(2) is completely broken while the vector U(2) gauge is left invariant. Note that the number and representations of the Higgs fields are fixed by our construction and are not at all arbitrary as in the usual treatment of the Higgs mechanism.

V. NATURE OF THE DYNAMICAL SYMMETRY BREAKDOWN

In the above I have revealed the fundamental equivalence underlying four-fermion and Yukawa theories. Basic steps of the argument were as follows.

(i) First a four-fermion Lagrangian is converted into a form,

$$L_F = L_f + L_{bf} + \text{(boson mass counter terms)}, \qquad (5\text{-}1)$$

by means of a functional technique. Here L_f is the fermion kinetic term and L_{bf} is the boson-fermion vertices. This theory is compared with

$$L_Y = L_f + L_{bf} + L_b + \text{(boson mass counter terms)}, \qquad (5\text{-}2)$$

where L_b is the boson kinetic and interaction vertices.

(ii) One-fermion-loop radiative correction is computed to L_F and L_Y. Radiatively corrected expressions are given by

$$\bar{L}_F = L_f + L_{bf} + CL_b + V_b, \tag{5-3}$$

$$\bar{L}_Y = L_f + L_{bf} + (1+C)L_b + V_b \ . \tag{5-4}$$

C is a certain divergent constant (in a symbolic notation) and V_b is the sum of n-boson vertices with $n>4$.

(iii)Renormalization of boson fields and vertices is performed differently in two theories,

$$C = \frac{1}{Z} \qquad \text{in } \bar{L}_F, \tag{5-5}$$

$$C = \frac{1}{Z} - 1 \quad \text{in } \bar{L}_Y, \tag{5-6}$$

and we obtain the same renormalized expression,

$$\bar{L} = L_f + L_{bf} + L_{b,R} + V_b \ . \tag{5-7}$$

Here the crucial point is that the radiative corrections are actually divergent, i.e., $C = \infty$, which makes it possible to eliminate the difference between C and 1+C by the redefinition of renormalization parameters. Since this difference amounts to the presence or absence of the independent bosonic degree of freedom in the original Lagrangian, we recognize that the ultraviolet divergence of the theory is the origin of the emergence of collective fields. If the ultraviolet divergence is not strong enough, C turns out to be finite and it is impossible to eliminate the difference between C and 1+C by renormalization convention. No bosonic fields are created in this case. Superconductivity models in two dimensions are these examples. Thus,

sufficiently strong ultraviolet divergence =>

emergence of collective fields.

On the other hand it is not difficult to realize that the origin of spontaneous symmetry breakdown lies in the infrared instability of the theory. Indeed in the Nambu-Jona-Lasinio model we had to introduce a vacuum value v_0 in order to avoid infrared divergences coming from the masslessness of the fermions. Shift of the σ field was the only way to obtain infrared finite results. Thus,

$v \neq 0 \leftarrow$ infrared instability.

Therefore the emergence of collective fields and the spontaneous symmetry breakdown have entirely different origins.

The appearance of a collective field does not necessarily imply the breakdown of its associated symmetry or vice versa. We exhibit a table below which summarizes the situation in typical field theoretic models.

Models	Emergence of fields	Symmetry Breakdown
Nambu-Jona-Lasinio	yes	yes (global γ_5)
Bjorken	yes	no
Gross-Neveu[8]	no	yes (global γ_5)
Coleman-Weinberg[9]	no	yes (gauge U(1))
ϕ^4	no	no

Finally we ask ourselves the question whether it is possible to obtain both the emergence of fields and symmetry breakdown within the framework of renormalizable models. At present this seems difficult since the appearance of collective fields entails the reduction of ultraviolet divergences of a theory. In the four-fermion model non-renormalizable interaction was reduced to a renormalizable one after bound state bosons were created. How is it possible to start with a renormalizable theory and to arrive at another renormalizable (not super-renormalizable) theory after new fields are created? This is an important problem which we have to answer in the near future.

At this workshop I came to know that Dr. N. Snyderman[10] has obtained similar results to mine in the Nambu-Jona-Lasinio model using a somewhat different method.

REFERENCES

1. W. Heisenberg, Z. Naturforsch. 14, 441 (1959).
2. Y. Nambu and G. Jona-Lasinio, Phys. Rev. 122, 345 (1961).
3. T. Eguchi, University of Chicago preprint EFI 76/20 (1976) and references cited therein.
4. J. Bjorken, Ann. Phys. 24, 174 (1963).
5. I. Bialynicki-Birula, Phys. Rev. 130, 465 (1963), G. S. Guralnik, Phys. Rev. 136, B1404 (1964).
6. T. Eguchi and H. Sugawara, Phys. Rev. D10, 4257 (1974).

7. A. Chakrabarti and B. Hu, Phys. Rev. D13, 2347 (1976).

8. D. Gross and A. Neveu, Phys. Rev. D10, 3235 (1975).

9. S. Coleman and E. Weinberg, Phys. Rev. D7, 1888 (1973).

10. N. Snyderman, in these proceedings.

Dynamically Broken Gauge Model Without Fundamental Scalar Fields

NEAL J. SNYDERMAN*
Institute for Theoretical Physics,
S.U.N.Y., Stonybrook, New York

and

G. S. GURALNIK†
Physics Department, Brown University
Providence, Rhode Island

ABSTRACT

It is shown that the structure that must be generated by dynamical symmetry breaking solutions to gauge theories can be explicitly implemented with a 4-fermion interaction. This structure arises in order to obtain consistency with the constraints imposed by a Goldstone commutator proportional to $<\bar{\psi}\psi>$. We demonstrate these ideas within the context of axial electrodynamics, dynamically breaking chiral symmetry. As a pre-requisite we show how the Nambu-Jona-Lasinio model becomes renormalizable with respect to a systematic approximation scheme that respects the Goldstone commutator of dynamically broken chiral symmetry to each order of approximation. (This approximation scheme is equivalent to a 1/N expansion, where N is set to unity at the end of the calculations.) This solution generates new interactions not explicitly present in the original Lagrangian and does not have a 4-fermion contact interaction. The renormalized Green's functions are shown to correspond to those of the σ-model, summed as though the fermions had N components, and for which $\lambda_0=2g_0^2$. This correspondence is exact except for the possibility that the renormalized coupling of the Nambu-Jona-Lasinio

*Supported by NSF Grant No. PHY-76-15328.

†Supported by ERDA Grant No.COO-3130TA-335, and Brown-HET-340.

model may be a determined number.

I. INTRODUCTION

Consider a gauge theory of Yang-Mills fields coupled to fermions, described by the Lagrangian

$$L = \bar{\psi}(i\not{\partial}-g\not{A}\cdot\underset{\sim}{T})\psi - \frac{1}{4}(\partial_\mu \underset{\sim}{A}_\nu - \partial_\nu \underset{\sim}{A}_\mu + g\underset{\sim}{A}_\mu \times \underset{\sim}{A}_\nu)^2 . \qquad (1.1)$$

During this conference we have heard about this theory within the context of the strong interactions where one is trying to understand how the infrared structure of the SU(3) color Yang-Mills fields is capable of binding the quarks into color singlet hadrons. A unified theory of the strong, weak and electromagnetic interactions would start with a large group that eventually breaks down to $SU(3)_{color}\times SU(2)\times U(1)_{weak+elect.}\rightarrow SU(3)_{color}\times U(1)_{elect}$. In this talk I would like to consider the possibility that a theory such as (1.1) is capable of realizing this unification by addressing the question of how a gauge theory without fundamental scalar fields can possess dynamically broken solutions.

We begin to consider this possibility by looking at the most general criteria for dynamically broken solutions. In order for a theory to possess broken symmetry solutions it must respect a Goldstone commutator.[1] An example of a Goldstone commutator for this theory is derived from the Green's function $<0|(j^a_\mu(z)\bar{\psi}(x)T^b\psi(x))_+|0>$, where j^a_μ are locally conserved currents. Taking the divergence and using the equal-time anticommutation relations for the fermion fields we obtain

$$\int d^4z\ \partial^z_\mu <0|(j^a_\mu(z)\bar{\psi}(x)T^b\psi(x))_+|0>$$

$$= <0|[Q^a,\bar{\psi}(x)T^b\psi(x)]|0> = -iC^{abc}\mathrm{Tr}\,T^c G. \qquad (1.2)$$

The condition $-i\mathrm{Tr}G=<0|\bar{\psi}\psi|0>\neq 0$ will imply that $Q^a|0>\neq 0$, the vacuum is not a null eigenstate of the charge, and that Q^a and $\bar{\psi}T^b\psi$ connect the vacuum to a particle state of the theory. Consistency of a solution to (1.1) with (1.2) will automatically maintain the Ward identity, but more importantly, it will imply the generation of new degrees of freedom not present in the original Lagrangian. These new degrees of freedom will enable gauge fields to become massive. There will also be scalar particles generated with self-couplings as well as couplings to gauge fields and fermions, which must also acquire mass. As only the fields in Eq. (1.1) carry dimensions, both the fermion and gauge field masses thus generated will be proportional to tr G.

One is immediately struck with the difficulty of implementing a constraint such as tr $G\neq 0$ since there is no tr G term in any of the Green's function equations of this theory, unlike models with fundamental scalar fields for which $<\phi>$ explicitly appears at the level of the field equations. For (1.1) ordinary perturbation theory cannot maintain the constraint of a non-vanishing Goldstone commutator proportional to trG; a non-perturbative lowest approximation is therefore necessary to implement the constraint. If such a non-perturbative lowest approximation is found, then since the theory with the resulting degrees of freedom is to describe a theory of electromagnetic coupling strength, we want to be able to perturb about this lowest approximation, consistently respecting the constraint of the Goldstone commutator order by order. We shall present below a model that realizes these features; it will have a non-perturbative lowest approximation which respects a Goldstone commutator with tr $G\neq 0$ and thus generates new degrees of freedom, and in which higher orders of approximation perturb about the lowest approximation in a way that is consistent with the Goldstone commutator to each order.

We introduce this model by considering axial electrodynamics of massless fermions described by the Lagrangian

$$L = \bar{\psi}(i\partial\!\!\!/-e_o A\!\!\!/\gamma_5)\psi- \frac{1}{4}(\partial_\mu A_\nu-\partial_\nu A_\mu)^2. \qquad (1.3)$$

If this model possesses solutions which dynamically break chiral symmetry[2], then the Goldstone commutator which must be respected is derived from the Green's function

$$<0|(j_{5_\mu}(z)i\bar{\psi}(x)\gamma_5\psi(x))_+|0>=i\int_{\xi,\eta} \mathrm{tr}\gamma_5 G(x,\xi)\Gamma_{5_\mu}(\xi,\eta,z)G(\eta,x). \qquad (1.4)$$

The relevant Goldstone commutator is

$$\int d^4z\partial^z_\mu<0|(j_{5_\mu}(z)i\bar{\psi}(x)\gamma_5\psi(x))_+|0>$$
$$= <0|[Q_5,i\bar{\psi}(x)\gamma_5\psi(x)]|0>= -2\mathrm{tr}G. \qquad (1.5)$$

With Eq. (1.4) we see that the Ward identity

$$i\partial^z_\mu\Gamma_{5_\mu}(\xi,\eta,z)=G^{-1}(\xi,z)\delta^4(z-\eta)\gamma_5+\gamma_5 G^{-1}(z,\eta)\delta^4(\xi-z) \qquad (1.6)$$

is a solution to Eq. (1.5) and therefore is satisfied whenever

(1.5) is. The implications of tr $G \neq 0$, though, will be far more dramatic than the reflection of chiral symmetry in the Green's functions implied by the Ward identity.

One of the manifestations of the breakdown of chiral symmetry of the vacuum, as a consequence of tr $G \neq 0$, will be a non-zero fermion mass with the proportionality

$$m_o \propto \text{tr } G. \tag{1.7}$$

In spontaneously broken theories with fundamental scalar fields, the fermion mass arises through the effective coupling in the Lagrangian

$$\bar{\psi}\psi\langle\phi\rangle, \tag{1.8}$$

which is here replaced by the effective coupling

$$\bar{\psi}\psi \text{ tr } G. \tag{1.9}$$

The non-perturbative lowest approximation to (1.3) must generate this structure. It is recognized that this kind of structure is associated with a 4-fermion interaction. We are therefore led to modify (1.3) by adding a 4-fermion interaction so that the structure of (1.9) can be made explicit. Thus we will consider the theory described by the chirally invariant Lagrangian[3]

$$L = \bar{\psi}(i\not{\partial}-e_o\not{A}\gamma_5)\psi - \frac{1}{4}(\partial_\mu A_\nu - \partial_\nu A_\mu)^2 + \frac{\lambda_o}{2}[(\bar{\psi}\psi)^2 + (i\bar{\psi}\gamma_5\psi)^2]. \tag{1.10}$$

We shall find that when the resulting theory is solved subject to the constraint of the Goldstone commutator (1.5) all of the symmetry breaking effects are explicitly generated in a systematic way, and that there is no 4-fermion contact interaction in the resulting theory!

II. NAMBU-JONA-LASINIO MODEL

As a prerequisite to studying this gauge model, let us first consider the model without the gauge field; this is the original model that introduced the ideas of spontaneous symmetry breaking in relativistic quantum field theory due to Nambu and Jona-Lasinio[4]. We are starting with a Lagrangian which is non-renormalizable with respect to perturbative solution in the coupling constant λ_o. However, not only is perturbation theory hopelessly divergent, but it is also incapable of respecting a constraint such as (1.5). When approximated in a way that respects (1.5) to each order of approximation, the model becomes renormalizable!

Such an approximation scheme is equivalent to the one obtained by considering N component fermion fields and summing the theory in the 1/N expansion, setting N=1 at the end of the calculations. This way of summing this theory is also appreciated by Eguchi[5]. We will obtain this approximation scheme, though, by using sources for the composite fields $\bar{\psi}\psi$ and $i\bar{\psi}\gamma_5\psi$.[6] The theory formulated this way, by Guralnik[7], has the advantage of making explicit the tr G structure, and leads in a natural way to the correct approximation scheme that respects the Goldstone commutator to each order of approximation. The Lagrangian for the Nambu-Jona-Lasinio model in the presence of sources is

$$L = \bar{\psi}i\partial\!\!\!/\psi + \frac{\lambda_o}{2}[(\bar{\psi}\psi)^2+(i\bar{\psi}\gamma_5\psi)^2]+J\bar{\psi}\psi + J_5\bar{\psi}\gamma_5\psi. \tag{2.1}$$

From the field equations the functional Schwinger-Dyson equation for

$$G(x,y) \equiv \frac{1}{i}\,\frac{\langle 0|(\psi(x)\bar{\psi}(y))_+|0\rangle}{\langle 0|0\rangle} \tag{2.2}$$

is obtained,

$$[i\partial\!\!\!/-\lambda_o i(\mathrm{tr}G(x,x)-\gamma_5\mathrm{tr}\gamma_5G(x,x))+J(x)+i\gamma_5J_5(x)$$
$$-\lambda_o i(\frac{\delta}{\delta J(x)} + i\gamma_5\,\frac{\delta}{\delta J_5(x)})]G(x,y) = \delta^4(x-y). \tag{2.3}$$

We see that the effect of the 4-fermion interaction is to generate the trG terms. This structure is obtained from the matrix element

$$\frac{\langle 0|(\bar{\psi}(x)\psi(x)\psi(x)\bar{\psi}(y))_+|0\rangle}{\langle 0|0\rangle} \tag{2.4}$$

that arises in the vacuum expectation value of the field equations after multiplying from the right with $\bar{\psi}(y)$ and time ordering. Using the Schwinger action principle, we can pull out the $\bar{\psi}\psi$ in Eq. (2.4) with the source J, obtaining

$$\langle 0|(\bar{\psi}(x)\psi(x)\psi(x)\bar{\psi}(y))_+|0\rangle=-i\,\frac{\delta}{\delta J(x)}\,\frac{\langle 0|(\psi(x)\bar{\psi}(y))_+|0\rangle}{\langle 0|0\rangle}$$
$$+\,\frac{\langle 0|\bar{\psi}(x)\psi(x)|0\rangle}{\langle 0|0\rangle}\,\frac{\langle 0|(\psi(x)\bar{\psi}(y))_+|0\rangle}{\langle 0|0\rangle}$$
$$=\frac{\delta}{\delta J(x)}\,G(x,y) + \mathrm{tr}G(x,x)G(x,y)\;. \tag{2.5}$$

Thus we see how this choice of sources generates the desired factorization of the matrix element (2.4) into (tr G) G+correction.

We approixmate the Schwinger-Dyson equation (2.3) by expanding G in a series in ε,

$$G = G^{(0)} + \varepsilon G^{(1)} + \varepsilon^2 G^{(2)} + \ldots , \tag{2.6}$$

while considering the functional derivative terms in (2.3) to be of order ε^8, thus

$$\begin{aligned}[i\not\partial - \lambda_o i(\mathrm{tr}\, G^{(0)} + \varepsilon\, \mathrm{tr}\, G^{(1)} + \varepsilon^2\, \mathrm{tr}\, G^{(2)} + \ldots) \\ + \lambda_o i\gamma_5(\mathrm{tr}\gamma_5 G^{(0)} + \varepsilon \mathrm{tr}\gamma_5 G^{(1)} + \ldots) + J + i\gamma_5 J_5 \\ - \lambda_o i\varepsilon(\frac{\delta}{\delta J} + i\gamma_5 \frac{\delta}{\delta J_5})](G^{(0)} + \varepsilon G^{(1)} + \ldots) = 1.\end{aligned} \tag{2.7}$$

Terms of the same order in ε are collected together, after which ε is set to unity. To zeroth order we have the equation

$$[i\not\partial - \lambda_o i(\mathrm{tr}G^{(0)} - \gamma_5 \mathrm{tr}\gamma_5 G^{(0)}) + J + i\gamma_5 J_5] G^{(0)} = 1, \tag{2.8}$$

and to order ε we have

$$\begin{aligned}G^{(0)-1} G^{(1)} - \lambda_o i\mathrm{tr}G^{(1)} G^{(0)} + \lambda_o i\gamma_5 \mathrm{tr}\gamma_5 G^{(1)} G^{(0)} \\ - \lambda_o i(\frac{\delta}{\delta J} + i\gamma_5 \frac{\delta}{\delta J_5}) G^{(0)} = 0,\end{aligned} \tag{2.9}$$

and so on.

The lowest approximation to the fermion propagator is obtained from Eq. (2.8) by turning off the sources. With the important identification

$$m_o \equiv \lambda_o i\; \mathrm{tr}G^{(0)}(x,x)\Big|_{J=J_5=0} \tag{2.10}$$

we obtain

$$G^{(0)}(x,y) = \int \frac{d^4p}{(2\pi)^4} e^{-ip(x-y)} \frac{1}{\not p - m_o + i\varepsilon} . \tag{2.11}$$

In order to obtain consistency between Eqs. (2.11) and (2.10) we are led to the constraint relation

$$1 = 4\lambda_o i \int \frac{d^4p}{(2\pi)^4} \frac{1}{p^2 - m_o^2 + i\varepsilon} . \tag{2.12}$$

The first order solution, obtained from Eq. (2.9) gives a perturbative correction to the fermion propagator,

$$G^{(1)}(x,y) = \lambda_o i \int_\xi G^{(0)}(x,\xi)\mathrm{tr}G^{(1)}(\xi,\xi)G^{(0)}(\xi,y) + \int_{\xi,w} G^{(0)}(x,\xi)\Sigma^{(1)}(\xi,w)G^{(0)}(w,y), \qquad (2.13)$$

where

$$\Sigma^{(1)}(\xi,w)=\lambda_o iG^{(0)}(\xi,w)\Delta^{(0)}(w,\xi)-\lambda_o i\gamma_5 G^{(0)}(\xi,w)\gamma_5\Delta_5^{(0)}(w,\xi), \qquad (2.14)$$

and where we have defined

$$\Delta^{(0)}(w,\xi) \equiv \lambda_o i\mathrm{tr}\frac{\delta G^{(0)}(w,w)}{\delta J(\xi)} - \delta^4(w-\xi) \qquad (2.15a)$$

and

$$\Delta_5^{(0)}(w,\xi) \equiv -\lambda_o \mathrm{tr}\gamma_5 \frac{\delta G^{(0)}(w,w)}{\delta J_5(\xi)} - \delta^4(w-\xi). \qquad (2.15b)$$

In arriving at these expressions we have used the fact that $GG^{-1}=1$ implies $\delta GG^{-1}+G\delta G^{-1}=0$ in order to express the variation of G in terms of the variation of G^{-1};the variation of $G^{(0)-1}$ is then obtained from Eq. (2.8). We want to make the identification of $\Delta^{(0)}$ and $\Delta_5^{(0)}$ with propagators for scalar and pseudoscalar particles respectively so that $\Sigma^{(1)}$ corresponds to ordinary self-energy radiative corrections to the fermions.

Consider Eq. (2.15b) for the pseudoscalar propagator. Evaluating the functional derivative using Eq. (2.8) we find the momentum space structure

$$\Delta_5^{(0)-1}(q) = -(1 + \lambda_o \Pi_5^{(0)}(q)), \qquad (2.16)$$

where

$$\Pi_5^{(0)}(q) = i\int \frac{d^4p}{(2\pi)^4} \mathrm{tr}\gamma_5 G^{(0)}(p)\gamma_5 G^{(0)}(p-q). \qquad (2.17)$$

$\Pi_5^{(0)}$ is a fermion loop with pseudoscalar vertices and is quadratically divergent. We isolate these divergences by expanding in a Taylor's series,

$$\Pi_5^{(0)}(q)=\Pi_5^{(0)}(o)+\frac{1}{2}q_\mu q_\nu \left.\frac{\partial^2\Pi_5^{(0)}(q)}{\partial q_\mu \partial q_\nu}\right|_0 +(\text{finite terms}). \qquad (2.18)$$

We recognize that

$$\lambda_o \Pi_5^{(0)}(o) = -4\lambda_o i \int \frac{d^4p}{(2\pi)^4} \frac{1}{p^2-m_o^2+i\varepsilon} = -1 \tag{2.19}$$

because of the constraint relation, Eq. (2.12). Therefore $\Delta_5^{-1}(q)$ goes like q^2, and thus describes a massless particle. This is the lowest approximation to the massless pseudoscalar particle implied by the non-vanishing Goldstone commutator, Eq. (1.5). The remaining evaluation gives the result

$$\Delta_5^{(0)}(q) = \frac{z^{-1}}{q^2\left[1-q^2 \frac{g_o^2}{8\pi^2} \int_{4m_o^2}^{\infty} \frac{d\kappa^2}{\kappa^2} \sqrt{1-\frac{4m_o^2}{\kappa^2}} \frac{1}{q^2-\kappa^2+i\varepsilon}\right]}, \tag{2.20}$$

where

$$z \equiv \frac{2\lambda_o}{i} \int \frac{d^4p}{(2\pi)^4} \frac{1}{(p^2-m_o^2+i\varepsilon)^2}, \tag{2.21}$$

and where we have defined the dimensionless bare coupling

$$g_o^2 \equiv \lambda_o/z. \tag{2.22}$$

Notice that this bare coupling is independent of λ_o, the original coupling of the Lagrangian.

The scalar propagator Eq. (2.15a) is analogously evaluated, with the result

$$\Delta^{(0)}(q) = \frac{z_s^{-1}}{(q^2-4m_o^2)\left[1-(q^2-4m_o^2)\frac{g_{s_o}^2}{8\pi^2} \int_{4m_o^2}^{\infty} \frac{d\kappa^2}{\kappa^2}\left(1-\frac{4m_o^2}{\kappa^2}\right)^{-\frac{1}{2}} \frac{1}{q^2-\kappa^2+i\varepsilon}\right]} \tag{2.23}$$

where

$$z_s = z\left[1+4m_o^2 \frac{g_o^2}{8\pi^2} \int_{4m_o^2}^{\infty} \frac{d\kappa^2}{\kappa^4}\left(1-\frac{4m_o^2}{\kappa^2}\right)^{-\frac{1}{2}}\right] \tag{2.24}$$

and where the scalar bare coupling is

$$g_{s_o}^2 \equiv \lambda_o/z_s . \tag{2.25}$$

To this order of approximation we then have for G,

$$G(x,y)= \int \frac{d^4p}{(2\pi)^4}\, e^{-ip(x-y)} \frac{1}{\not{p}-\lambda_o i(\mathrm{tr}G^{(0)}+\mathrm{tr}G^{(1)})-\sum^{(1)}(p)}\,, \tag{2.26}$$

where we have extracted the self energy parts from Eq. (2.13). The fermion mass is defined by the pole,

$$m - \lambda_o i(\mathrm{tr}G^{(0)} + \mathrm{tr}G^{(1)}) - \sum^{(1)}(m) = 0, \tag{2.27}$$

so

$$G^{-1}(p) = \not{p}-m-(\sum^{(1)}(p) - \sum^{(1)}(m)). \tag{2.28}$$

From the radiative corrections to $G^{(0)}$ from $\Delta^{(0)}$ and $\Delta_5^{(0)}$ we absorb the logarithmic divergence into Z_2,

$$\sum^{(1)}(p) - \sum^{(1)}(m) = (\not{p}-m)(1-Z_2^{-1}) + \bar{\sum}^{(1)}(p), \tag{2.29}$$

where $\bar{\Sigma}^{(1)}(p)$ is finite. Therefore, Eq. (2.28) is an approximation to the structure

$$G(p) = \frac{Z_2}{\not{p}-m-\bar{\bar{\sum}}(p)+ i\varepsilon}\,. \tag{2.30}$$

We next consider the lowest approximation to the Goldstone commutator to demonstrate that it is respected. The Green's function

$$<0|(j_{5_\mu}(z)i\bar{\psi}(x)\gamma_5\psi(x))_+|0>=-\mathrm{tr}\gamma_\mu\gamma_5 \frac{\delta G(z,z)}{\delta J_5(x)}\,, \tag{2.31}$$

is calculated in lowest approximation from Eq. (2.8) for $G^{(0)}$ giving

$$-i\int_\xi \mathrm{tr}\gamma_\mu\gamma_5 G^{(0)}(z,\xi)\gamma_5 G^{(0)}(\xi,z)\Delta_5^{(0)}(\xi,x) = - \frac{2m_o}{\lambda_o}\int \frac{d^4q}{(2\pi)^4}\, e^{-iq(z-x)} \frac{q_\mu}{q^2+i\varepsilon}\,. \tag{2.32}$$

Substituting Eq. (2.32) into Eq. (1.5) we find, to this order of approximation,

$$<0|[Q_5, i\bar{\psi}(x)\gamma_5\psi(x)]|0> = - \frac{2m_o}{\lambda_o i} = -2\mathrm{tr}G^{(0)}, \tag{2.33}$$

verifying the consistency of our calculations.

The lowest approximation answers we have obtained for $G^{(0)}$, $\Delta^{(0)}$, and $\Delta_5^{(0)}$ reproduce the original results of Nambu and Jona-Lasinio. Their resummation of the perturbation series is seen to be the one that is consistent with the Goldstone commutator to lowest order. We have seen how the lowest order fermion propagator is corrected by the scalar and pseudoscalar propagators, and we will next consider corrections to the bound-state propagators.

The functional expressions Eqs. (2.15) for the bound-state propagators in terms of the fermion propagator can be easily seen to be lowest approximations to expressions that are valid to all orders. From our Schwinger-Dyson equation for the fermion propagator, Eq. (2.3), we can make a natural identification of scalar and pseudoscalar fields[7] (vacuum expectation values in the presence of sources),

$$\phi(x) \equiv \lambda_o i \operatorname{tr} G(x,x) - J(x), \tag{2.34a}$$

and

$$\phi_5(x) \equiv -\lambda_o \operatorname{tr}\gamma_5 G(x,x) - J_5(x). \tag{2.34b}$$

From these fields we define the propagators[7]

$$\Delta(x,y) \equiv \frac{\delta\phi(x)}{\delta J(y)} = \lambda_o i \operatorname{tr} \frac{\delta G(x,x)}{\delta J(y)} - \delta^4(x-y), \tag{2.35a}$$

and

$$\Delta_5(x,y) \equiv \frac{\delta\phi_5(x)}{\delta J_5(y)} = -\lambda_o \operatorname{tr}\gamma_5 \frac{\delta G(x,x)}{\delta J_5(y)} - \delta^4(x-y). \tag{2.35b}$$

That these are the correct expressions to all orders also emerges from higher order corrections to the fermion propagator.

We can now obtain the next correction to $\Delta_5^{(0)}$ from

$$\Delta_5^{(1)}(x,y) = -\lambda_o \operatorname{tr}\gamma_5 \frac{\delta G^{(1)}(x,x)}{\delta J_5(y)} . \tag{2.36}$$

From Eq. (2.9) in the presence of sources we can evaluate the functional derivative and find the perturbative structure

$$\Delta_5^{(1)}(x,y) = \lambda_o \int_{\xi,w} \Delta_5^{(0)}(x,\xi)\Pi_5^{(1)}(\xi,w)\Delta_5^{(0)}(w,y). \tag{2.37}$$

Extracting the polarization correction $\Pi_5^{(1)}$, the pseudoscalar propagator to this order of approximation has the momentum space structure

$$\Delta_5(q) = \frac{-1}{1 + \lambda_o(\Pi_5^{(0)}(q) + \Pi_5^{(1)}(q))} . \tag{2.38}$$

The polarization $\Pi_5^{(1)}(q)$ has the graphical structure shown in Fig. (1). Evaluating these integrals at q=0, we find the very important result that the leading divergences cancel, that is

$$\Pi_5^{(1)}(0) = 0 ! \tag{2.39}$$

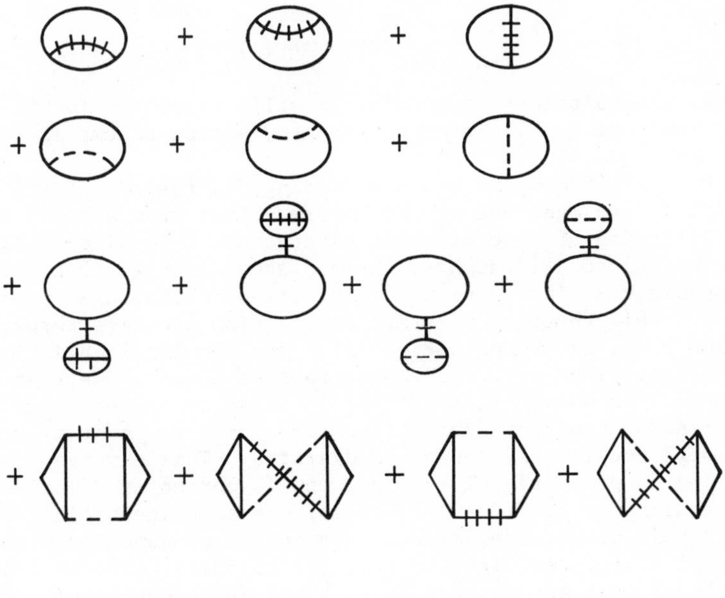

FIGURE 1. Graphs contributing to $\Pi_5^{(1)}(q)$. The lines represent the following propagators:

—— $G^{(0)}$, ┼┼┼┼┼ $\Delta^{(0)}$, - - - - - $\Delta_5^{(0)}$

Referring to Eq. (2.38) for $\Delta_5(q)$ and recalling Eq. (2.19), we see that (2.39) implies that the pseudoscalar remains massless. The cancellations required for $\Pi_5^{(1)}(0)=0$ occur between the "2-loop" and "3-loop" diagrams. All "3-loop" diagrams have an extra logarithmic divergence,

$$\frac{2}{i}\int \frac{d^4p}{(2\pi)^4}\frac{1}{(p^2-m_o^2+i\varepsilon)^2}, \tag{2.40}$$

which combines with the extra factor of λ_o to give z (recall Eq. (2.21)). This factor of z cancels the extra factor of z^{-1} from the extra bound-state propagator (recall Eqs. (2.20),(2.23) and (2.24)).

The masslessness of Δ_5 to this order of approximation combined with the non-vanishing of tr G to this order of approximation,

$$\mathrm{tr}G = \mathrm{tr}G^{(0)}(1 + C^{(1)}) \neq 0, \tag{2.41}$$

implies the Goldstone commutator is still respected to this order of approximation. Therefore our approximation scheme is consistent with $\partial_\mu j_{5\mu}=0$.

We next consider the non-leading divergences of the graphs in Fig. (1) to show how new structure arises from a consistent identification of some of these divergences. Consider a diagram contributing to $\Pi_5^{(1)}$ such as the one shown in Fig. (2), which has subdiagrams involving the interaction of the three bound-states. This three bound-state interaction proceeds through a fermion loop, as shown in Fig. (3). The fermion loop is logarithmically divergent, and we subtract it at q=0. We then have two terms, as shown in Fig. (4), a finite subtracted fermion loop, and a divergent subtraction term which we identify with a bare self coupling of three bound states. This divergent bare coupling is proportional to $m_o g_o$, and is correctly identified by the requirement $\Pi_5^{(1)}(0)=0$. (That is, the same identification of logarithmic divergences discussed above in conjunction with the integral (2.40), required so that $\Pi_5^{(1)}(0)=0$, is also made here). The theory thus generates a three bound state coupling! The diagram of $\Pi_5^{(1)}$ we have been considering in Fig. (2) therefore corresponds to the set of diagrams shown in Fig. (5). Similarly, we can also obtain new structure by looking at the ordinary radiative corrections to the pseudoscalar vacuum polarization, such as the graph in Fig. (6). There is a logarithmic divergence associated with the subdiagram of four bound states interacting through a fermion loop. Again, subtracting this divergence of q=0 gives a bare coupling of four bound states as the subtraction

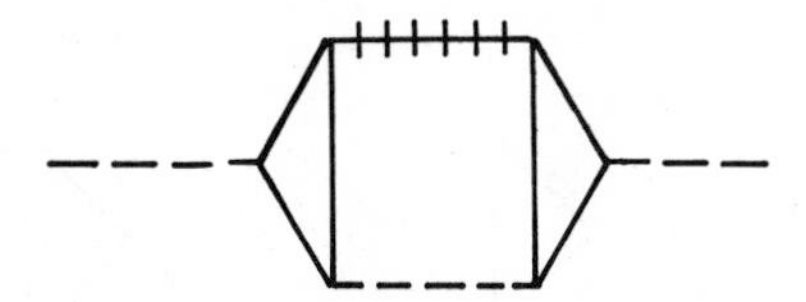

FIGURE 2. Contribution to $\Pi_5{}^{(1)}$ with subdiagrams involving the interaction of three bound-states. (External lines are explicitly shown.)

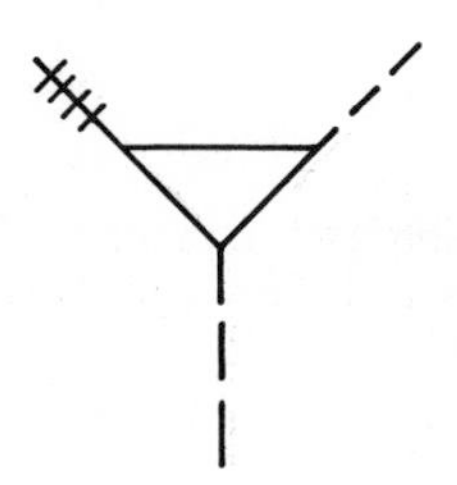

FIGURE 3. Three bound-state interaction through a fermion loop.

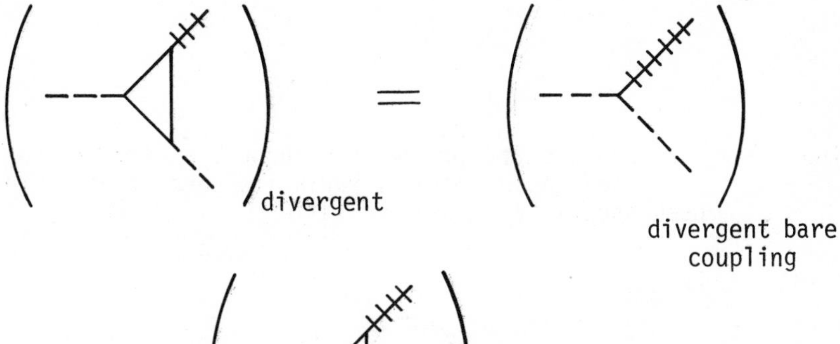

divergent

=

divergent bare coupling

+

finite subtracted fermion loop

FIGURE 4. Subtraction of the fermion triangle loop with the identification of a bare three bound-state coupling.

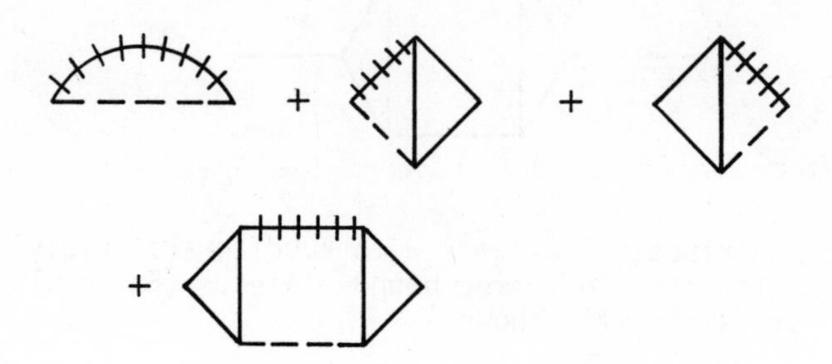

FIGURE 5. Equivalent graphical content of polarization graph of Fig. (2). Here all fermion triangle loops are subtracted and finite.

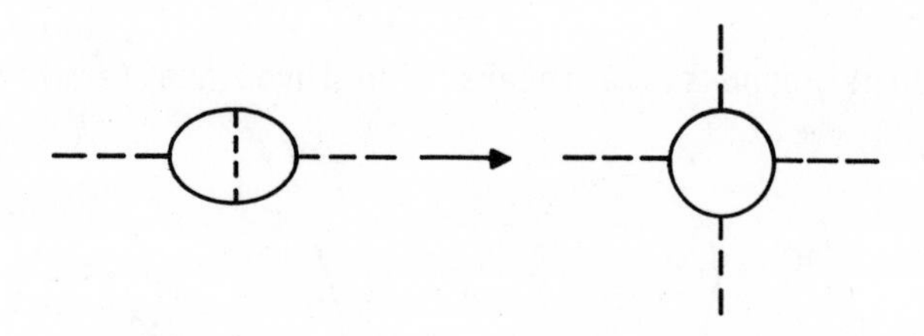

FIGURE 6. Four bound-state interaction through a fermion loop contained in an ordinary radiative correction to the pseudoscalar vacuum polarization.

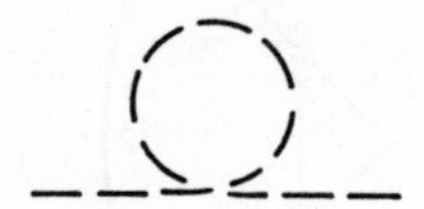

FIGURE 7. Polarization contribution associated with subtraction term of four bound state interaction of Fig. (6).

term. This coupling is proportional to g_o^2. The new polarization contribution associated with this coupling is shown in Fig. (7). Thus through a consistent identification of divergences in the graphs of Figs. (3) and (6) for the interaction of three and four bound states through fermion loops new vertices are generated. There are no higher point couplings, though, since fermion loops with more vertices are convergent.

The graphs contributing to $\Pi_5^{(1)}$ group into the fermion self-energy corrections we have seen before, and into pseudoscalar-fermion vertex corrections. The subsequent renormalization of these functions, and the renormalization of Δ_5 leads to a renormalized pseudoscalar-fermion coupling. The original dimensional coupling λ_o was renormalized away in the lowest approximation to $g_o^2 \equiv \lambda_o/z$, a dimensionless bare coupling. This bare coupling is now renormalized to

$$g^2 \equiv g_o^2 \, Z_3 \, \frac{Z_2^2}{Z_1^2} \, . \tag{2.42}$$

In ordinary renormalizable theories the bare coupling can be chosen to diverge in a way that cancels the divergence of the Z's leaving an arbitrary renormalized coupling. Here, however, the bare coupling g_o^2 has a fixed divergence structure implying the possibility that if this divergence cancels with that of the Z's, the renormalized coupling g^2 would be a determined number!

We have seen the generation of new interactions not explicitly present in the original Lagrangian. We now show there is no renormalized counterpart to λ_o by showing there is no longer a four-fermion contact interaction in the solution. The amplitude of four fermion fields

$$K(x,y,z,w) \equiv \left(\frac{1}{i}\right)^2 \frac{\langle 0|(\psi(x)\bar{\psi}(y)\psi(z)\bar{\psi}(w))_+|0\rangle}{\langle 0|0\rangle} \, , \tag{2.43}$$

satisfies the equation

$$\begin{aligned} &[i\partial\!\!\!/ - \lambda_o i(\mathrm{tr}G(x,x) - \gamma_5 \mathrm{tr}\gamma_5 \, G(x,x)) + J(x) + i\gamma_5 J_5(x) \\ &- \lambda_o i(\frac{\delta}{\delta J(x)} + i\gamma_5 \, \frac{\delta}{\delta J_5(x)})]K(x,y,z,w) \\ &= \delta^4(x-y)G(z,w) - \delta^4(x-w)G(z,y). \end{aligned} \tag{2.44}$$

This equation is similar to the one for G, except that the right-hand side goes like the difference of G's (this incorporates the exchange interaction required by the Pauli principle). It is approximated in an analogous way, expanding G in a series in ε as before, with the functional derivative of order ε, and also expanding K,

$$K = K^{(0)} + \varepsilon K^{(1)} + \varepsilon^2 K^{(2)} + \ldots \quad . \tag{2.45}$$

In the lowest approximation there is only free propagation of $G^{(0)}$'s. The lowest order scattering contributions come from $K^{(1)}$ and have the structure

$$\begin{aligned} &\lambda_o i \int_{\xi,\eta} [G^{(0)}(x,\xi)G^{(0)}(\xi,y)]\Delta^{(0)}(\eta,\xi)[G^{(0)}(z,\eta)G^{(0)}(\eta,w)] \\ &-\lambda_o i \int_{\xi,\eta} [G^{(0)}(x,\xi)\gamma_5 G^{(0)}(\xi,y)]\Delta_5^{(0)}(\eta,\xi)[G^{(0)}(z,\eta)\gamma_5 G^{(0)}(\eta,w)] \\ &+ \text{exchange interaction.} \end{aligned} \tag{2.46}$$

The interaction between the fermions is mediated by the bound-states. It is recalled from the definitions of the bound-state propagators, Eqs. (2.15), that there are δ-function terms which, in an expansion in λ_o would give rise to the 4-fermion contact interaction. However, these terms are cancelled in the evaluation of the bound-state propagators. The symmetry breaking consistency condition, Eq. (2.12) that creates the bound-states cancels the 1 in these propagators, thus eliminating any chance of a 4-fermion contact interaction. Also, recalling the z^{-1} in the expressions for $\Delta^{(0)}$ and $\Delta_5^{(0)}$, Eqs. (2.20) and (2.23), we see that the λ_o's in Eq. (2.46) are transformed away, and so in no sense is there a renormalized version of λ_o which measures the strength of a 4-fermion interaction.

We have found a systematic approximation scheme for this model that is consistent with the Goldstone commutator to each order of approximation, and which reproduces the original results of Nambu and Jona-Lasinio as a lowest approximation. The symmetry and resulting particle and interaction content of our solution is the same as that of the σ-model. We will now demonstrate the connection between these models.

III. CONNECTION WITH σ-MODEL

The σ-model is described by the Lagrangian

$$L = \bar{\psi}[i\not{\partial}-g_o(\sigma+i\gamma_5\pi)]\psi + \frac{1}{2}[(\partial_\mu\sigma)^2+(\partial_\mu\pi)^2]$$
$$- \frac{1}{2}\mu_o^2(\sigma^2+\pi^2)-\frac{\lambda_o}{4}(\sigma^2+\pi^2) - J_\sigma\sigma - J_\pi\pi, \qquad (3.1)$$

where we have added sources for the σ and π fields. The σ and π fields transform like the fermion operators

$$\begin{aligned} \sigma &: \bar{\psi}\psi \\ \pi &: i\bar{\psi}\gamma_5\psi. \end{aligned} \qquad (3.2)$$

From the field equations derived from (3.1) we calculate the equation for the vacuum expectation values in the presence of sources,

$$\phi_a(x) \equiv \frac{<0|\begin{pmatrix}\sigma(x)\\ \pi(x)\end{pmatrix}|0>}{<0|0>}, \qquad (3.3)$$

with the result

$$(-\partial^2-\mu_o^2)\phi_a(x)+\lambda_o \mathrm{tr}\,\frac{\delta\Delta(x,x)}{\delta J_a(x)} - \lambda_o i\Delta_{ab}(x,x)\phi_b(x)$$
$$-\lambda_o i\phi_b(x)\Delta_{ba}(x,x)-\lambda_o i\mathrm{tr}\Delta(x,x)\phi_a(x)-\lambda_o\phi^2(x)\phi_a(x)$$
$$+ig_o\mathrm{tr}G(x,x)\delta_{a\sigma}-g_o\mathrm{tr}\gamma_5 G(x,x)\delta_{a\pi}=J_a(x), \qquad (3.4)$$

where we have defined the propagators

$$\Delta_{ab}(x,y) \equiv \delta\phi_a(x)/\delta J_b(y), \qquad (3.5)$$

and where the fermion propagator satisfies

$$\left[i\not{\partial}-g_o(\phi_\sigma(x)+i\gamma_5\phi_\pi(x))-ig_o\left(\frac{\delta}{\delta J_\sigma(x)}+i\gamma_5\frac{\delta}{\delta J_\pi(x)}\right)\right]G(x,y)$$
$$= \delta^4(x-y). \qquad (3.6)$$

We approximate these exact equations by considering Δ of order ε, as well as all explicit functional derivatives. The lowest approximation to Eq. (3.4) for the σ-component thus becomes

$$(-\partial^2-\mu_o^2)\phi_\sigma^{(0)}(x)-\lambda_o\phi^{(0)2}(x)\phi_\sigma^{(0)}(x)+ig_o\mathrm{tr}G^{(0)}(x,x)=J_\sigma(x). \qquad (3.7)$$

This lowest approximation differs from the usual one by the inclusion of the tr G term. As we have mentioned that our summation of the Nambu-Jona-Lasinio model is equivalent to a 1/N expansion, the inclusion of tr G in lowest approximation here is what one would obtain for N component fermion fields. With the sources turned off we get the relation for the vacuum expectation value

$$\eta \equiv \phi_\sigma^{(0)}\Big|_{J=0}, \qquad (3.8)$$

$$(\mu_o^2+\lambda_o\eta^2)\eta = ig_o\mathrm{tr}\, G^{(0)}. \qquad (3.9)$$

In the usual solution where the tr $G^{(0)}$ term is neglected in lowest approximation, there are two solutions: either $\eta=0$ (normal solution), or $\eta^2=-\mu_o^2/\lambda_o$ (spontaneously broken solution) which requires the bare mass to be chosen imaginary. Now from the lowest approximation to Eq. (3.6) for G we obtain

$$[i\not\partial-g_o(\phi_\sigma^{(0)}(x)+i\gamma_5\phi_\pi^{(0)}(x))]G^{(0)}(x,y)=\delta^4(x-y). \qquad (3.10)$$

With the sources turned off and with the identification

$$m_o \equiv g_o\eta, \qquad (3.11)$$

we obtain

$$G^{(0)}(x,y) = \int \frac{d^4p}{(2\pi)^4}\, e^{-ip(x-y)}\, \frac{1}{\not p-m_o+i\varepsilon}. \qquad (3.12)$$

With this expression for $G^{(0)}$, our Eq. (3.9) for η gives the constraint equation

$$1 = \frac{4g_o^2 i}{\mu_o^2+\lambda_o\eta^2}\int \frac{d^4p}{(2\pi)^4}\, \frac{1}{p^2-m_o^2+i\varepsilon}. \qquad (3.13)$$

This equation will play the analogous role in this solution to the σ-model as Eq. (2.12) in the Nambu-Jona-Lasinio model.

We can now calculate the lowest approximation to the π-propagator. We begin with Eq. (3.4) which in lowest approximation gives for the π-component,

$$(-\partial^2-\mu_o^2)\phi_\pi^{(0)}(x)-\lambda_o\phi^{(0)2}(x)\phi_\pi^{(0)}(x)-g_o\mathrm{tr}\gamma_5G^{(0)}(x,x)=J_\pi(x). \qquad (3.14)$$

The equation for Δ_π is obtained by functionally differentiating this equation with respect to J_π. This leads to the equation

$$[q^2-\mu_o^2-\lambda_o\eta^2-g_o i\int\frac{d^4p}{(2\pi)^4}\,\mathrm{tr}\gamma_5 G^{(0)}(p)\gamma_5 G^{(0)}(p-q)]\Delta_\pi^{(0)}(q)=1, \tag{3.15}$$

where we have evaluated the functional derivative $\mathrm{tr}\gamma_5\delta G^{(0)}(x,x)/\delta J_5(y)$ with Eq. (3.10), then turned off the sources and Fourier transformed. The fermion loop gives

$$-g_o^2 i\int\frac{d^4p}{(2\pi)^4}\,\mathrm{tr}\gamma_5 G^{(0)}(p)\gamma_5 G^{(0)}(p-q)=4g_o^2 i\int\frac{d^4p}{(2\pi)^4}\frac{1}{p^2-m_o^2+i\varepsilon}$$
$$+\text{ terms of order } q^2. \tag{3.16}$$

Using the constraint Eq. (3.13), we see that Eq. (3.15) describes the propagation of a massless particle. Completing the evaluation of the integrals we obtain

$$\Delta_\pi^{(0)}(q)=\frac{Z_\pi}{q^2\left[1-q^2\,\frac{g^2}{8\pi^2}\int_{4m_o^2}^{\infty}\frac{d\kappa^2}{\kappa^2}\sqrt{1-\frac{4m_o^2}{\kappa^2}}\,\frac{1}{q^2-\kappa^2+i\varepsilon}\right]}, \tag{3.17}$$

where now

$$1-Z_\pi^{-1}\equiv 2g_o^2 i\int\frac{d^4p}{(2\pi)^4}\,\frac{1}{(p^2-m_o^2+i\varepsilon)^2}, \tag{3.18}$$

and

$$g^2\equiv g_o^2\,Z_\pi\,. \tag{3.19}$$

Comparison of this expression for $\Delta_\pi^{(0)}$ with that for $\Delta_5^{(0)}$, Eq. (2.20), in the Nambu-Jona-Lasinio model shows that they differ only in the replacements $Z_\pi\leftrightarrow z^{-1}$ and $g^2=g_o^2 Z_\pi\leftrightarrow g_o^2=\lambda_o/z$. While the unrenormalized propagators differ in this way, the renormalized ones will be functionally identical. There is one possible difference, though, in that the renormalized coupling of the Nambu-Jona-Lasinio model may be a determined number, while in the σ-model it is an arbitrary parameter.

Next we shall investigate the σ-propagator. Functionally differentiating Eq. (3.7), with respect to J_σ leads to the equation

$$\left[q^2-\mu_o^2-3\lambda_o\eta^2+ig_o^2\int\frac{d^4p}{(2\pi)^4}\,\mathrm{tr}G^{(0)}(p)G^{(0)}(p-q)\right]\Delta_\sigma^{(0)}(q)=1. \tag{3.20}$$

The evaluation of the fermion loop gives

$$ig_o^2\int\frac{d^4p}{(2\pi)^4}\,\mathrm{tr}G^{(0)}(p)G^{(0)}(p-q)=4g_o^2 i\int\frac{d^4p}{(2\pi)^4}\frac{1}{p^2-m_o^2+i\varepsilon}$$
$$+\text{ terms of order }(q^2-4m_o^2). \tag{3.21}$$

Thus Eq. (3.20) becomes,using the constraint Eq. (3.13),

$$[q^2-2\lambda_o\eta^2-(q^2-4m_o^2)\text{terms}]\Delta_\sigma^{(0)}(q) = 1. \tag{3.22}$$

Now the scalar propagator in the Nambu-Jona-Lasinio model has a mass

$$m_\sigma^2 = 4m_o^2\ , \tag{3.23}$$

so in order to obtain a correspondence between these models we must make the restriction

$$\lambda_o = 2g_o^2\ . \tag{3.24}$$

In this case $\Delta_\sigma^{(0)}$ evaluates to

$$\Delta_\sigma^{(0)}(q)=\frac{Z_\sigma}{(q^2-4m_o^2)\left[1-(q^2-4m_o^2)\frac{g_\sigma^2}{8\pi^2}\int_{4m_o^2}^{\infty}\frac{d\kappa^2}{\kappa^2}\left(1-\frac{4m_o^2}{\kappa^2}\right)^{-\frac{1}{2}}\frac{1}{q^2-\kappa^2+i\varepsilon}\right]}\ , \tag{3.25}$$

where

$$Z_\sigma^{-1} = Z_\pi^{-1}\left[1+\frac{4m_o^2 g^2}{8\pi^2}\int_{4m_o^2}^{\infty}\frac{d\kappa^2}{\kappa^4}\left(1-\frac{4m_o^2}{\kappa^2}\right)^{-\frac{1}{2}}\right]\ , \tag{3.26}$$

and where

$$g_\sigma^2 = g_o^2 Z_\sigma. \tag{3.27}$$

This expression for $\Delta_\sigma^{(0)}$ corresponds to the one for $\Delta^{(0)}$ in the Nambu-Jona-Lasinio model, Eq. (3.23), with the replacements $Z_\sigma \leftrightarrow z_s^{-1}$, and $g_\sigma^2=g_o^2Z_\sigma\leftrightarrow g_{so}^2=\lambda_o/z_s$. Again, while the unrenormalized propagators differ in this way the renormalized ones are functionally identical.

We can also consider the next corrections to the π-propagator. The resulting polarization $\Pi_\pi^{(1)}$ has the graphical structure of Fig. (8). We explicitly see in this model all the graphs

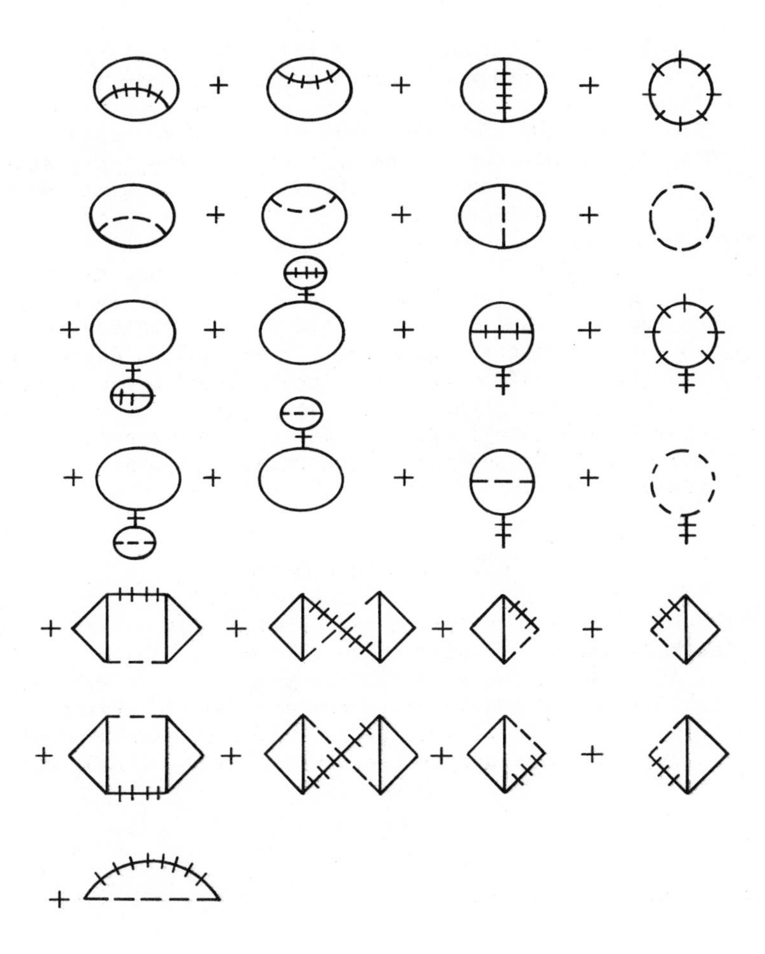

FIGURE 8. Graphs contributing to $\Pi_\pi^{(1)}(q)$. The lines now represent ——— $G^{(0)}$, —┼┼┼┼┼— $\Delta_\sigma^{(0)}$, — — — — — $\Delta_\pi^{(0)}$.

that were generated in the Nambu-Jona-Lasinio model by a proper identification of the divergences of subdiagrams. The divergent subdiagrams in the σ-model must be similarly subtracted in the renormalization process, but the logarithmic divergences are now identified with $1-Z_\pi^{-1}$ through Eq. (3.18). The resulting renormalized Δ_π to this order of approximation is again functionally identical to the corresponding renormalized Δ_5 of the Nambu-Jona-Lasinio model! We also note that if we had not made the restriction $\lambda_o=2g_o^2$ in our lowest approximation to the σ-propagator, needed to gain correspondence with the Nambu-Jona-Lasinio model, then we would not be able to satisfy the Goldstone theorem in this way of summing the σ-model.

We are thus led to the startling conclusion that two entirely different Lagrangians, one non-renormalizable in perturbation theory and the other renormalizable in perturbation theory, when solved subject to the constraint of the same Goldstone commutator, Eq. (1.5), lead to the same renormalized Green's functions (up to the possibility that the coupling in the Nambu-Jona-Lasinio model is a number)! We consider this correspondence a kind of "universality" associated with respecting a given Goldstone commutator.

IV. AXIAL ELECTRODYNAMICS

With this universality principle in mind, let us return to axial electrodynamics. We will solve this model subject to the constraint of dynamically broken chiral symmetry. We will be able to explicitly and consistently respect the constraint of the Goldstone commutator, Eq. (1.5), by including the Nambu-Jona-Lasinio interaction. We add to the Lagrangian Eq. (1.10) the source terms

$$L' = J\bar{\psi}\psi + J_5 i\bar{\psi}\gamma_5\psi + J_\mu A_\mu . \tag{4.1}$$

With this choice of sources we derive the Schwinger-Dyson equations

$$\begin{aligned}[i\partial\!\!\!/ - e_o A\!\!\!/(x)\gamma_5 + ie_o\gamma_\mu\gamma_5 \frac{\delta}{\delta J_\mu(x)} - \lambda_o i(\mathrm{tr}G(x,x) - \gamma_5 \mathrm{tr}\gamma_5 G(x,x))\\ + J(x) + i\gamma_5 J_5(x) - \lambda_o i(\frac{\delta}{\delta J(x)} + i\gamma_5 \frac{\delta}{\delta J_5(x)})]G(x,y) = \delta^4(x-y),\end{aligned} \tag{4.2}$$

where the equations for the vacuum expectation value of the gauge field in the presence of sources

$$a_\mu(x) \equiv \langle 0|A_\mu(x)|0\rangle/\langle 0|0\rangle \tag{4.3}$$

become, in the radiation gauge,

$$a_o(x) = \frac{1}{\nabla^2}\,[ie_o \mathrm{tr}\gamma_o\gamma_5 G(x,x) + J_o(x)], \tag{4.4}$$

and

$$-\partial^2 a_K^T + ie_o \mathrm{tr}\gamma_K^T\gamma_5 G(x,x) = J_K^T(x). \tag{4.5}$$

If the terms proportional to λ_o in Eq. (4.2) are ignored, then these equations are just those of ordinary axial electrodynamics in the presence of sources. The new terms due to the inclusion of the 4-fermion interaction produce the explicit tr G terms in Eq. (4.2). The evalution of functional derivatives of a_μ will now give quite different results from ordinary axial electrodynamics because of the additional terms in the equation for G. Because of these terms, we shall be able to consistently generate all the structure implied by the non-vanishing Goldstone commutator.

Once again we approximate these equations by expanding G and a_μ in a series in ε and considering all functional derivatives of order ε. The lowest approximation for G again corresponds to dropping the functional derivative terms, and with the sources turned off we again obtain a free massive fermion propagator, with the identification $m_o=\lambda_o i\ \mathrm{tr}\ G^{(0)}$. We also obtain the consequent constraint, Eq. (2.12). In the next approximation we obtain the correction to $G^{(0)}$ with the same structure as Eq. (2.13), but now

$$\begin{aligned}
\Sigma^{(1)}(\xi,w) = {} & \lambda_o iG^{(0)}(\xi,w)\left[\lambda_o i\ \mathrm{tr}\ \frac{\delta G^{(0)}(\xi,\xi)}{\delta J(w)} - \delta^4(\xi-w)\right] \\
& -\lambda_o i\gamma_5 G^{(0)}(\xi,w)\gamma_5\left[-\lambda_o \mathrm{tr}\gamma_5\ \frac{\delta G^{(0)}(\xi,\xi)}{\delta J_5(w)} - \delta^4(\xi-w)\right] \\
& -e_o^2\gamma_5 G^{(0)}(\xi,w)\gamma_o\gamma_5\lambda_o i\ \frac{1}{\nabla_\xi^2}\ \mathrm{tr}\gamma_o\gamma_5\ \frac{\delta G^{(0)}(\xi,\xi)}{\delta J_5(w)} \\
& -ie_o\gamma_o\gamma_5 G^{(0)}(\xi,w)\gamma_5\lambda_o i\ \mathrm{tr}\gamma_5\ \frac{\delta G^{(0)}(\xi,\xi)}{\delta J_o(w)} \\
& -ie_o^2\gamma_o\gamma_5 G^{(0)}(\xi,w)\gamma_o\gamma_5\frac{1}{\nabla_\xi^2}\left[ie_o \mathrm{tr}\gamma_o\gamma_5\ \frac{\delta G^{(0)}(\xi,\xi)}{\delta J_o(w)} - \delta^4(\xi-w)\right] \\
& +ie_o\gamma_K^T\gamma_5 G^{(0)}(\xi,w)\gamma_\ell^T\gamma_5\ \frac{\delta a_\ell^{(0)T}(\xi)}{\delta J_K^T(w)}\ .
\end{aligned} \tag{4.6}$$

As in obtaining Eqs. (2.14) and (2.15) for $\Sigma^{(1)}$ in the Nambu-Jona-Lasinio model, these terms arise from expressing the functional derivatives of $G^{(0)}$ in terms of $G^{(0)-1}$, where from Eqs. (4.2) and (4.4),

$$G^{(0)-1}(\xi,\eta)=\Big[i\not\partial_\xi-\lambda_o i(\mathrm{tr}G^{(0)}(\xi,\xi)-\gamma_5\mathrm{tr}\gamma_5G^{(0)}(\xi,\xi))$$
$$-e_o\gamma_o\gamma_5\frac{1}{\nabla_\xi^2}(ie_o\mathrm{tr}\gamma_o\gamma_5G^{(0)}(\xi,\xi)+J_o(\xi))$$
$$+e_o\gamma_K^T\gamma_5a_K^{(0)T}(\xi)+J(\xi)+i\gamma_5J_5(\xi)\Big]\delta^4(\xi-\eta). \quad (4.7)$$

We see that there are many degrees of freedom contributing to the radiative corrections to the fermions. If we were to turn off λ_o, the last two terms would be the only ones to contribute to $\Sigma^{(1)}$ in Eq. (4.6), and these terms would give the perturbation correction to the fermions due to an axial photon in the radiation gauge. Now, however, the evaluation of the functional derivatives gives very different results.

As an example, consider the term that would have given only a Coulomb interaction if the λ_o terms were not present,

$$\Delta_{oo}^{(0)}(\xi,w)=\frac{\delta a_o^{(0)}(\xi)}{\delta J_o(w)}=\frac{1}{\nabla_\xi^2}\Big[ie_o\mathrm{tr}\gamma_o\gamma_5\frac{\delta G^{(0)}(\xi,\xi)}{\delta J_o(w)}+\delta^4(\xi-w)\Big]. \quad (4.8)$$

From Eq. (4.7) for $G^{(0)-1}$ we calculate

$$\Delta_{oo}^{(0)}(\xi,w)=\frac{1}{\nabla_\xi^2}\Big[\delta^4(\xi-w)-$$
$$-\frac{ie_o^2}{\nabla_w^2}\int_{\rho,\sigma}\mathrm{tr}\gamma_o\gamma_5G^{(0)}(\xi,\rho)\Gamma_{5_o}^{(0)}(\rho,\sigma;w)G^{(0)}(\sigma,\xi)\Big]^{-1} \quad (4.9)$$

where

$$\Gamma_{5_o}^{(0)}(\rho,\sigma;w)\equiv\gamma_o\gamma_5\delta^4(\rho-\sigma)\delta^4(\rho-w)$$
$$+\lambda_o i\gamma_5\int_u\Delta_5^{(0)}(\rho,u)\mathrm{tr}\gamma_5G^{(0)}(u,w)\gamma_o\gamma_5G^{(0)}(w,u)\delta^4(\rho-\sigma). \quad (4.10)$$

This is the time component of the vertex associated with a conserved axial current, and evaluates in momentum space to

$$\Gamma^{(0)}_{5_0}(q) = \gamma_0\gamma_5 - 2m_o \frac{q_o}{q^2}\gamma_5 \ . \tag{4.11}$$

The Fourier transform of Eq. (4.9) has the structure

$$\Delta^{(0)}_{00}(q) = \frac{-1/\underset{\sim}{q}^2}{1 - e_o^2/\underset{\sim}{q}^2\, \Pi^{(0)}_{5_{00}}(q)} \ . \tag{4.12}$$

The axial vector polarization can be expressed in terms of the vacuum polarization of quantum electrodynamics,

$$e_o^2\Pi^{(0)}_{5_{00}}(q) = e_o^2\Pi^{(0)}_{00}(q) + \frac{\mu_o^2}{z}\frac{\underset{\sim}{q}^2}{q^2}\frac{\Delta_5^{(0)-1}(q)}{q^2} ,$$

where

$$\mu_o^2 \equiv \frac{4e_o^2 m_o^2}{g_o^2} ,$$

and where g_o^2 is again the dimensionless bare coupling λ_o/z. Equation (4.12) then becomes

$$\Delta^{(0)}_{00}(q) = \frac{(1 - q_o^2/\underset{\sim}{q}^2)}{q^2 + e_o^2 q^2 \Pi^{(0)}_{(q)} - \left(\mu_o^2/z\right)\Delta_5^{(0)-1}(q)/q^2} \ . \tag{4.13}$$

Performing the renormalization subtractions on the spectral representations for the vacuum polarization $\Pi^{(0)}_{(q)}$ and the pseudoscalar propagator $\Delta_5^{(0)}(q)$, we obtain the result

$$\Delta^{(0)}_{00}(q) =$$

$$\frac{(1 - q_o^2/\underset{\sim}{q}^2)\ Z_{3M}}{(q^2 - M_o^2 + i\varepsilon)\left[1 - (q^2 - M_o^2)\dfrac{e_M^2}{12\pi^2}\displaystyle\int_{4m_o^2}^{\infty}\frac{d\kappa^2}{\kappa^2}\frac{\left(1 - \dfrac{4m_o^2}{\kappa^2}\right)^{3/2}}{q^2 - \kappa^2 + i\varepsilon}\ P\frac{1}{\left(1 - \dfrac{M_o^2}{\kappa^2}\right)^2}\right]} \tag{4.14}$$

which has a radiation gauge numerator and the denominator of a massive particle. The gauge field mass M_0^2, which corresponds to the pole in Eq. (4.13), is to a very crude approximation

$$M_o^2 \sim \mu_o^2. \tag{4.15}$$

On considering the other components contributing to the fermion self-energy in Eq. (4.6) we find that, except for the first term which represents the exchange of a scalar particle with the same propagator as Eq. (2.23), all other terms are proportional to

$$D(q) \equiv \frac{Z_{3M}}{(q^2-M_o^2+i\varepsilon)\left[1-(q^2-M_o^2)\dfrac{e_M^2}{12\pi^2}\displaystyle\int_{4m_o^2}^{\infty}\frac{d\kappa^2}{\kappa^2}\frac{\left(1-\dfrac{4m_o^2}{\kappa^2}\right)^{3/2}}{q^2-\kappa^2+i\varepsilon}\;P\,\frac{1}{\left(1-\dfrac{M_o^2}{\kappa^2}\right)^2}\right]}, \tag{4.16}$$

and thus all these terms contribute to the propagation of a massive gauge field.

We next verify that we have produced a massive spin-1 particle by considering the gauge invariant fermion scattering amplitude mediated by this particle. The 4-fermion amplitude

$$K(x,y,z,w)=\left(\frac{1}{i}\right)^2\frac{\langle 0|(\psi(x)\bar\psi(y)\psi(z)\bar\psi(w))_+|0\rangle}{\langle 0|0\rangle} \tag{4.17}$$

now satisfies the radiation gauge equation

$$\begin{aligned}&\Big[i\not{\partial}-ie_o^2\gamma_0\gamma_5\frac{1}{\nabla^2}\mathrm{tr}\gamma_0\gamma_5G(x,x)-e_o\gamma_0\gamma_5\frac{1}{\nabla^2}J_0(x)\\&+e_o\gamma_K^T\gamma_5a_K^T(x)-\lambda_oi(\mathrm{tr}G(x,x)-\gamma_5\mathrm{tr}\gamma_5G(x,x))\\&+J(x)+i\gamma_5J_5(x)-\lambda_oi\Big(\frac{\delta}{\delta J(x)}+i\gamma_5\frac{\delta}{\delta J_5(x)}\Big)+ie_o\gamma_0\gamma_5\frac{\delta}{\delta J_0(x)}\\&+ie_o\gamma_K^T\gamma_5\frac{\delta}{\delta J_K^T(x)}\Big]K(x,y,z,w)\\&=\delta^4(x-y)G(z,w)-\delta^4(x-w)G(z,y).\end{aligned} \tag{4.18}$$

From the lowest order scattering contribution to this Green's function we can calculate the on-shell scattering amplitude by reduction. The various non-covariant terms contributing to the fermion self-energy in Eq. (4.6) here combine to produce the covariant and gauge invariant exchange,

$$\frac{ie_M^2\,\bar{u}\,\gamma_\mu\gamma_5\,u\left(g_{\mu\nu}-\frac{q_\mu\,q_\nu}{M_o^2\,F(q^2)}\right)\bar{u}\,\gamma_\nu\,\gamma_5\,u}{(q^2-M_o^2+i\varepsilon)\left[1-(q^2-M_o^2)\,\frac{e_M^2}{12\pi^2}\int_{4m_o^2}^{\infty}\frac{d\kappa^2}{\kappa^2}\,\frac{\left(1-\frac{4m_o^2}{\kappa^2}\right)^{3/2}}{q^2-\kappa^2+i\varepsilon}\;P\,\frac{1}{\left(1-\frac{M_o^2}{\kappa^2}\right)^2}\right]}. \tag{4.19}$$

The numerator of this propagator satisfies

$$\left.\left(-g_{\mu\nu}+\frac{q_\mu\,q_\nu}{M^2\,F(q^2)}\right)\right|_{q^2=M_o^2}=-g_{\mu\nu}+\frac{q_\mu q_\nu}{M_o^2}$$

$$=\sum_{\substack{3\text{ polarization}\\\text{states}}}\varepsilon_\mu(q)\varepsilon_\nu(q), \tag{4.20}$$

demonstrating that a massive spin-1 particle has in fact been created.

Also contributing to the scattering is the exchange of a massive scalar particle. Its coupling to fermions is proportional to $e_o m_o/M_o$ (since $g_o^2=4e_o^2m_o^2/(4e_o^2m_o^2/g_o^2)=4e_o^2m_o^2/\mu_o^2$). Just as the importance of the scalar was felt in the Nambu-Jona-Lasinio model in radiative corrections to the pseudoscalar vacuum polarization, its importance will be felt in the gauge theory when higher orders of approximation are considered, in which case diagrams such as those shown in Fig. (9) will be necessary for the renormalizability of the theory[12].

We now demonstrate that our lowest approximation with all the newly generated structure respects the Ward identity. The lowest approximation to the Green's function

$$\langle 0|(j_{5\mu}(z)\,i\bar{\psi}(x)\gamma_5\psi(x))_+|0\rangle$$

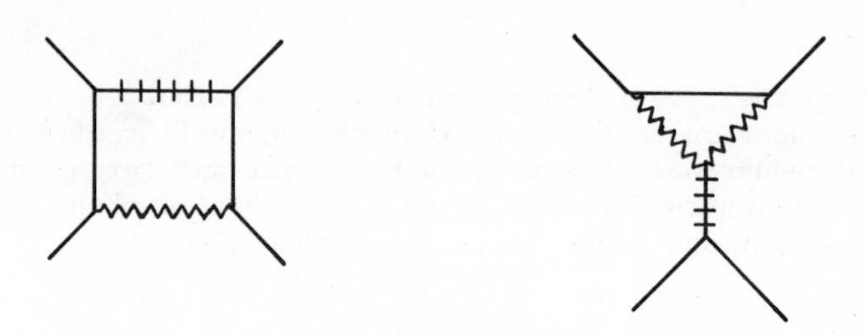

FIGURE 9. Examples of graphs involving scalars in fermion scattering that play an important role in the renormalizability to this order of approximation. The line ~~~~ represents the sum of terms contributing to the propagation of the massive gauge field.

gives the beautiful radiation gauge result,

$$<0|(j_{5\mu}(z)i\bar{\psi}(x)\gamma_5\psi(x))_+|0> = -\frac{\delta}{\delta J_5(x)} \operatorname{tr}\gamma_\mu\gamma_5 G(z,z)$$

$$= \int \frac{d^4q}{(2\pi)^4} e^{-iq(z-x)} -2\operatorname{tr}G^{(0)}\Big[e_o^2\Pi_{\mu o}^{(0)}(q)\frac{q_o}{q^2}\Delta_{00}^{(0)}(q)$$

$$+\mu_o^2/z\, g_{\mu o}\left(\Delta_5^{(0)-1}(q)/q^2\right)\frac{q_o}{q^2}\Delta_{00}^{(0)}(q) + \frac{q_\mu}{q^2}\Delta_5^{(0)-1}(q)\Delta_{55}^{(0)}(q)\Big], \tag{4.21}$$

where $\Delta_{55}{}^{(0)}(q)$ is the lowest approximation to the "would-be Goldstone boson", that is, it has the same functional expression as Δ_5 without the gauge field. Due to the presence of the additional structure in Eq. (4.7) for $G^{(0)-1}$, we find

$$\Delta_{55}^{(0)}(\xi,w) = -\lambda_o \operatorname{tr}\gamma_5 \frac{\delta G^{(0)}(\xi,\xi)}{\delta J_5(w)} - \delta^4(\xi-w)$$

$$= \int \frac{d^4q}{(2\pi)^4} e^{-iq(\xi-w)} \frac{\mu_o^2}{z}\left[\frac{1}{M_o^2 F(q^2)} + \frac{1}{\underset{\sim}{q}^2}\right] D(q^2), \tag{4.22}$$

where $D(q^2)$ was given in Eq. (4.16) and where $F(q^2)$ has the property that $F(M_o^2)=1$. We can also express $\Delta_{55}^{(0)}(q)$ in terms of $\Delta_{00}^{(0)}(q)$ through the relation

$$\Delta_5^{(0)-1}(q)\Delta_{55}^{(0)}(q) = 1-\frac{\mu_o^2}{z}\frac{q_o^2}{q^2}\left(\Delta_5^{(0)-1}(q)/q^2\right)\Delta_{00}^{(0)}(q), \tag{4.23}$$

with which we can easily calculate from Eq. (4.21),

$$<0|[Q_5, i\bar{\psi}(x)\gamma_5\psi(x)]|0> = \int d^4z \ \partial^z_\mu <0|(j_{5\mu}(z) i\bar{\psi}(x)\gamma_5\psi(x))_+|0>$$

$$= -2\mathrm{tr}\ G^{(0)}. \qquad (4.24)$$

Therefore, the equal-time Goldstone commutator is satisfied, and consequently so is the Ward identity. But what about the Goldstone theorem?

To understand why the Goldstone commutator Eq. (4.24) does not imply a massless Goldstone boson we investigate the Guralnik, Hagen and Kibble theorem[9]. We shall see that even though the equal-time Goldstone commutator implies $\partial_\mu j_{5\mu} = 0$, Q_5 actually becomes time dependent, which circumvents the Goldstone theorem. To see this we make use of the discontinuity formula

$$<0|[A(x), B(y)]|0> = 2i\,\mathrm{Im}\,\varepsilon(x_o - y_o)<0|(A(x), B(y))_+|0>. \qquad (4.25)$$

Applying this to

$$\int d^3z <0|(j_{50}(z)\ i\ \bar{\psi}(x)\gamma_5\psi(x))_+|0>, \qquad (4.26)$$

from Eq. (4.21) we find

$$<0|[Q_5(z_o),\ i\ \bar{\psi}(x)\gamma_5\psi(x)]|0>$$
$$= 2i\ \mathrm{Im}\int d^3z\ \varepsilon(z_o - x_o)<0|(j_{50}(z) i\bar{\psi}(x)\gamma_5\psi(x))_+|0>$$
$$= -\frac{2m_o}{\lambda_o}\int \frac{d^4q}{(2\pi)^4}\ e^{-iq(z-x)}\ q_o\ D(q^2)$$
$$= -\ 2\mathrm{tr}\ G^{(0)}\ \cos M_o(z_o - x_o)\ . \qquad (4.27)$$

Thus we see that the global charge has become time dependent. We can see how this is consistent with the local conservation of the axial current by considering

$$\frac{d}{dz_o} <0|[Q_5(z_o), i\bar{\psi}(x)\gamma_5\psi(x)]|0>$$
$$= \int d^3z <0|[\partial^z_o j_{50}(z), i\bar{\psi}(x)\gamma_5\psi(x)]|0>$$
$$= \int d^3z <0|[(\partial^z_\mu j_{5\mu}(z) - \nabla_{\underset{\sim}{z}} \underset{\sim}{j}_5(z)), i\bar{\psi}(x)\gamma_5\psi(x)]|0>$$
$$= -\int d\underset{\sim}{\sigma}^z <0|[\underset{\sim}{j}_5(z), i\bar{\psi}(x)\gamma_5\psi(x)]|0>. \qquad (4.28)$$

The time dependence of the Goldstone commutator requires that current is flowing through the surface at infinity. In his analysis of the circumvention of the Goldstone theorem, Higgs[10] showed how the loss of manifest Lorentz invariance in the radiation gauge would violate the Goldstone consequence of spontaneous symmetry breaking. We explicitly see from Eq. (4.28) that this loss of manifest Lorentz invariance enables the right hand side to be non-zero; however, it is the charge oscillation which makes explicit the preservation of the equal-time Goldstone commutator and thus the Ward identity.

V. CONCLUSION

We have constructed a dynamically broken solution to axial electrodynamics by introducing the Nambu-Jona-Lasinio 4-fermion interaction in order to explicitly produce the tr G structure which must emerge in order to respect the equal-time Goldstone commutator. We have seen that without the gauge field, the solution of the remaining model corresponds to the σ-model. We attribute this correspondence to a universality associated with respecting the same Goldstone commutator. Without the gauge field there is a Goldstone theorem that accompanies consistency with the Goldstone commutator, and in some sense this theorem is dictating the solution to these theories. In the gauge theory, on the other hand, we have seen that there is no longer a Goldstone theorem in the radiation gauge. However in the Lorentz gauge, to lowest approximation we have

$$\langle 0|(j_{5_\mu}(z)i\bar{\psi}(x)\gamma_5\psi(x))_+|0\rangle = \int \frac{d^4q}{(2\pi)^4}\, e^{-iq(z-x)}\, \frac{q_\mu}{q^2}\,, \qquad (5.1)$$

which implies that the Goldstone commutator is time independent, so consequently the Goldstone theorem is still valid. The resulting Goldstone boson is a gauge mode, though. Since the S-matrix is gauge invariant, even with a gauge field the Goldstone commutator still dictates correlations in the solution to the theory, but now it will demand massive gauge fields. We see that even for a gauge theory, the Goldstone commutator is as great a constraint on the solution as in theories without gauge fields.

Because of this central importance of the Goldstone commutator on the solution to a theory, we believe that if axial electrodynamics possesses a non-perturbative solution that dynamically breaks chiral symmetry, then the solution must have functionally identical Green's functions to the ones we have found by introducing an additional 4-fermion interaction to explicitly implement the constraint of the Goldstone commutator. We believe that this universality will also apply to the Yang-Mills gauge theory. It is clearly a restriction on the freedom of gauge theory modeling to consider that the symmetry of a fermion multiplet (as opposed to that of an explicit scalar multiplet) is to be dynamically broken[11]. Also, it is recalled that our correspondence between the Nambu-Jona-Lasinio model and the σ-model is exact up to the possibility that, without the explicit scalar fields, the Nambu-Jona-Lasinio model is more constrained in a way that offers it the potential to determine its renormalized coupling as a number. This is suggestive of the attractive possibility that if gauge theories without fundamental scalar fields can dynamically break down, then a unified gauge theory of weak and electromagnetic interactions may determine the couplings and masses of all particles involved in terms of α and m_e.

Note Added in Proof. Some of our conclusions regarding the renormalization of the Nambu-Jona-Lasinio model and its correspondence with the σ-model have also been reached by T. Eguchi, (Univ. of Chicago preprint), from another point of view.

ACKNOWLEDGMENT

One of us (N.S.) would like to thank Professor T.W.B. Kibble for his interest and his hospitality at Imperial College while this report was being prepared. We also wish to thank Dr. S. Kasdan for many enlightening discussions.

REFERENCES

* Details of all of the topics discussed in this talk are contained in Neal J. Snyderman, Ph.D. Thesis, Brown University, May 1976 (unpublished).

1. J. Goldstone, A. Salam and S. Weinberg, Phys. Rev. 127, 965 (1962).

2. This possibility has been studied by R. Jackiw and K. Johnson, Phys. Rev. D8, 2386 (1973), from the point of view of F. Englert and R. Brout, Phys. Rev. Lett. 13, 321 (1964).

3. This model has also been investigated by Y. Freundlich and D. Lurie, Nucl. Phys. B19, 557 (1970).

4. Y. Nambu and G. Jona-Lasinio, Phys. Rev. 122, 345 (1961).

5. T. Eguchi, Talk given at this Conference. The 2-D version of the Nambu-Jona-Lasinio model has been studied using the 1/N expansion by D. Gross and A. Neveu, Phys. Rev. D10, 3235 (1974).

6. The correspondence between the 1/N expansion and functional approximation schemes using sources for composite fields has been studied within the context of scalar theories by F. Cooper, G. S. Guralnik and S. Kasdan, Phys. Rev. (to be published). See also Gross and Neveu, footnote 5.

7. G. S. Guralnik, Nuovo Cimento 36, 1002 (1965).

8. This approximation scheme was used by T. F. Wong and G. S. Guralnik, Phys. Rev. D3, 3028 (1971) in calculating higher loop approximations to the σ-model.

9. G. S. Guralnik, C. R. Hagen and T.W.B. Kibble, Phys. Rev. Lett. 13, 585 (1964).

10. P. W. Higgs, Phys. Lett. 12, 132 (1964).

11. It is natural to consider mechanisms involving dynamical mass splitting as emphasized by S. Kasdan, Ph.D. Thesis, Brown University, June 1976, (unpublished), and talk given at this conference. A fermion model analogous to his scalar model is presently being studied. See also J. M. Cornwall and R. E. Norton, Phys. Rev. D8, 3338 (1973).

12. In the next order of approximation we generate a triangle anomaly (S. L. Adler, Phys. Rev. 177, 2426 (1969); J. S. Bell and R. Jackiw, Nuovo Cimento 60A, 47 (1969)) associated with a fermion loop with three axial-vector vertices. Since the conserved axial current is of central importance in our discussion, we rotate the effects of the anomaly on to the vector current, preventing its conservation. Since we do not couple to the vector current, the anomaly is not relevant to the aspects of the problem we wish to stress.

Hydrodynamic Model of Collective Resonances in Hadronic Matter

VICTOR A. MATVEEV*
Fermi National Accelerator Laboratory,
Batavia, Illinois

ABSTRACT

We study the collective resonance phenomena in multiquark hadronic systems. Our analysis is essentially qualitative and based on an analogy to the well-known giant resonance phenomena in the nuclear matter.

We consider the non-relativistic hydrodynamic equations for a two-component compressible fluid describing a system of quarks and antiquarks confined to the interior of the finite volume. The confinement properties as well as the relevant phase transitions are discussed and the frequencies of the hydrodynamic oscillations of the system are derived.

The problem of taking into account non-Abelian colored gluon fields in the hydrodynamic equations describing motion of confined quark/antiquark fluids ("chromo-hydrodynamics") is briefly discussed and the estimation of the effect of quark-gluon interactions on the energies of collective resonances is given.

We speculate that the collective resonances - or as we call them the hadronic giant resonances - may play an essential role in understanding the new resonances observed recently in e^+e^- -annihilation and lepton pair production experiments in unexpectedly high mass intervals. Particularly, we show that the energies of the lowest hadronic giant resonances which could be seen in e^+e^- annihilation (E0 and E2 modes) lie at $\sim$4 GeV if a size of the confinement region is about 1 GeV^{-1} ($\sim$0.2 fm). We should expect also the existence of colored collective states with energies of about the same order as for uncolored ones, which cannot be seen however in the process of e^+e^- annihilation into hadrons due to the color conservation in

*Permanent address: Joint Institute for Nuclear Research, Dubna, USSR.

strong interactions.

I. INTRODUCTION

The hypothesis on the existence of the collective resonance phenomena in hadronic systems which could be the analogy to the giant resonances in nuclei has been introduced in paper (1).

The main motive was the unexpected and intriguing at that time results of the experimental studies of e^+e^- -annihilation into hadrons which have shown the relatively large and growing with energy yield of hadrons. While contradicting the predictions of the old-fashioned naive parton model,these results and the following discovery of the ψ/J-particles have stimulated an invention of new quantum numbers and corresponding new degrees of freedom (charm, color etc.). The authors of the paper[1] argued that for a consistent explanation of the new events it might be useful to take into account an excitation of collective degrees of freedom beyond the elementary ones which mainly are treated by the parton model.

Considering, in particular, quark-antiquark pairs as weakly coupled quasiparticles of hadronic matter produced in e^+e^- -annihilation, one may expect collective excitations to appear with energies depending on the size of the hadronic system, the radius of interaction between constituents,etc.

Starting from the simplest variant of a model of hydrodynamic oscillations of two incompressible fluids, one for quarks and another for antiquarks (compare with the Goldhaber-Teller model of the dipole giant resonances in nuclei[2]), the authors of paper[1] derived a qualitative estimation of energies of collective hadronic resonances and pointed to some feasible implications.

The recent exciting discovery of the narrow resonance $\gamma(6.0)$ in the e^+e^- -production experiment[3] as well as the observation of the rich resonance "mini-structure" at $\sim$4 GeV in e^+e^--annihilation into hadrons[4] give us reason to renew the discussion of the collective resonance phenomena.

In this paper I shall study the hydrodynamic equations for the two-component compressible fluid describing a superdense system of quarks and antiquarks confined to the interior of a finite volume. Due to the assumption that masses of free unconfined quarks are very large, the consideration is mainly non-relativistic, and an analogy to the theory of the nuclear giant resonance[5] plays an essential role.

So the analogy of the "symmetry energy" potential, well-known in nuclear physics, which describes in our case fluctuations of energy near the point of the phase transition between the symmetric phase (all the local densities of quarks and antiquarks coincide) and the antisymmetric one (the local densities of quarks and antiquarks do not coincide) is introduced.

We argue that the confined symmetric phase is stable under fluctuations of the total quark-antiquark local density $\rho=\rho_q+\rho_{\bar{q}}$, so the corresponding fluid can be considered as the incompressible one.

The frequencies of the hydrodynamic oscillations in the confined hadronic phase are derived and the estimation of the energies of the corresponding "hadronic giant resonances" (Γ-series) is done, bearing in mind that these resonances may give a possible explanation of the "mini-structure" in the total cross-section of e^+e^--annihilation.

Then we discuss the problem of incorporation of the gluon degrees of freedom in the framework of the hydrodynamic approach. The first approximation to the so-called "chromodynamics", describing motion of the quark-antiquark fluids in the presence of colored gluon fields, is derived, and an estimation is given of the effect of quark-gluon interactions on the frequencies of collective resonances.

II. GIANT RESONANCE PHENOMENA IN NUCLEAR AND HADRONIC PHYSICS

The giant resonance phenomena in nuclear photoabsorption reactions give a good example of collective excitations in systems of strongly interacting particles. Although the properties of hadronic matter composed of confined colored quarks and gluons should be different from those of nuclear matter, we believe that the physics underlying the collective resonance phenomena in nuclear and hadronic matter is quite similar. For this reason, we begin first with a short introduction to the nuclear giant resonance physics.

It was known for a long time that cross sections of inelastic photoabsorption reactions on nuclei (when measured with sufficiently bad resolution) develop broad peaks--named the giant resonances--with positions slightly dependent on nucleus mass number somewhere between 13 and 25 MeV.[5] (See Fig. 1.) The rather big widths of the giant resonances as well as the strength of the absorption are considered as an indication of the collective nature of this effect, which cannot be explained by an excitation of single-particle degrees of freedom alone.

The giant electric dipole resonance was described by Goldhaber and Teller as a collective oscillation of the protons as a whole against the neutrons as a whole in the nucleus.[2] So the large absorption of γ-quanta with an electric field coupled to the dipole operator

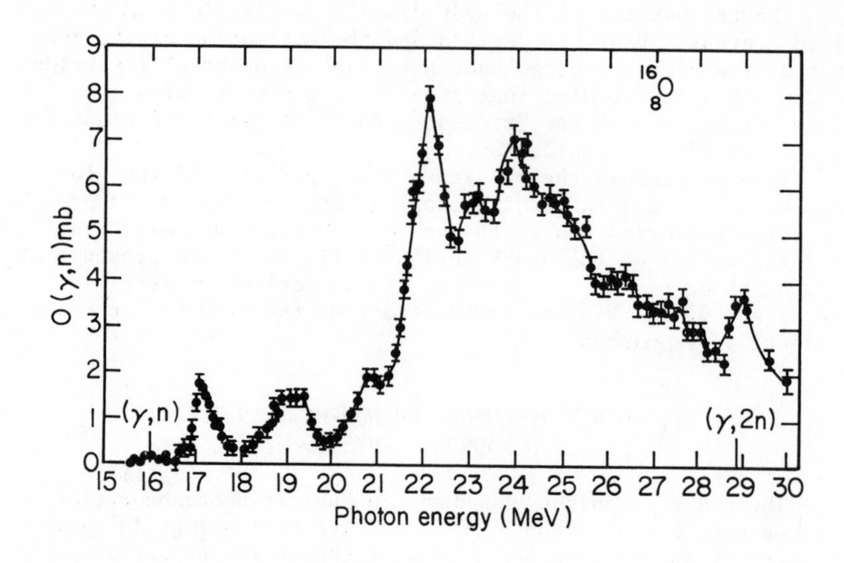

FIGURE 1. The photoneutron cross section of O^{16} up to 30 MeV, [taken from Bramblett, Caldwell, Harvey, Fultz[20]].

$$\vec{D} = \sum_i e_i \vec{r}_i \Rightarrow \frac{ZN}{A} (\vec{r}_p - \vec{r}_n) , \tag{2.1}$$

where $(\vec{r}_p - \vec{r}_n)$ is the relative center of mass coordinate vector of the protons and the neutrons, appears at photon energies determined by the frequencies of collective oscillations (E1, I=1 mode).

A necessary requirement for an excitation of collective resonances is the so-called coherence condition,

$$\lambda\!\!\!^{-} \gg R , \tag{2.2}$$

where $\lambda\!\!\!^{-}$ is a wave vector of the electric field, and $R \sim A^{1/3}$ is the size of the nucleus.

Assuming that protons and neutrons inside the nucleus can be considered as hard interpenetrating spheres and that the total restoring force is proportional to the nuclear surface, Goldhaber and Teller derived the formula which determines the frequencies of collective oscillations

$$\omega \approx \left(\frac{3u_0}{Rr_0 M_p} \right)^{1/2} \approx \frac{35 \text{ MeV}}{A^{1/6}} . \tag{2.3}$$

Here u_0 is an "ionization" potential of nuclear matter and r_0 the radius of the nuclear forces.

Obviously there exist, in general, besides the dipole (E1) oscillations the others, e.g. the monopole (E0), the quadrupole (E2), etc. oscillations, as illustrated in Fig. 2.

Moreover, theories of nuclear matter show that collective oscillations can deal with all degrees of freedom such as spin and isotopic spin.[6] So, there exist modes of vibrations in which protons with spin up and neutrons with spin down move against protons with spin down and neutrons with spin up (spin-isospin mode), or in which nucleons with spin up move against nucleons with spin down (spin-wave mode) etc. (see illustration in Fig. 3). If one considers the excitation of basic 0^+ states of nuclei, this leads to Wigner's supermultiplets of giant resonances which are degenerate under the assumption of spin-isospin independence of nuclear forces.[7]

An analysis of the giant resonance phenomena has shown that their most adequate description is given by a consideration of the classical motion of a two-fluid system which was initially applied to the nuclear problem by A. B. Migdal (1945)[8] and by H. Steinwedel, et al. (1950).[9] The important idea of the hydrodynamical approach was that the restoring force is described by the potential energy density

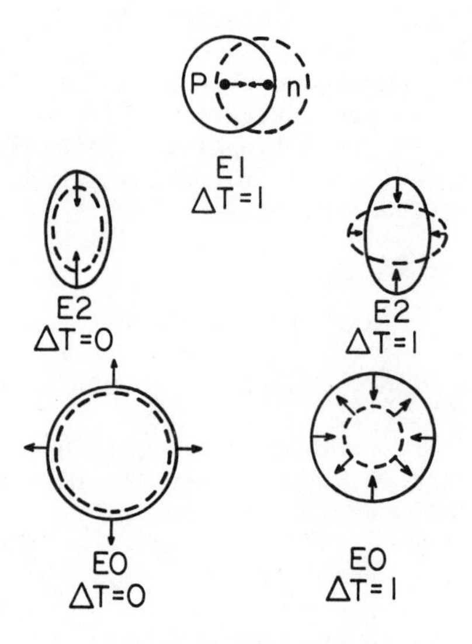

FIGURE 2. A schematic representation of giant resonances as hydrodynamic oscillations.

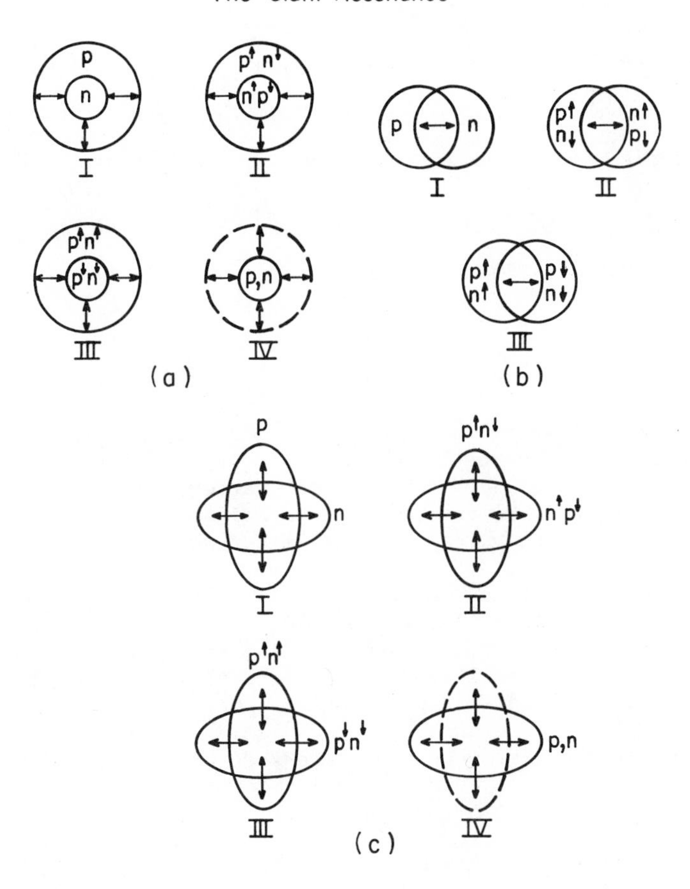

FIGURE 3. Scheme of collective multipole vibrations of nuclear matter when spin (s) and isospin (i) are taken into account (modes of the generalized Goldhaber-Teller model): (a) E0: (I) i mode, (II)si, (III)s, (IV)protons and neutrons in phase; (b) E1: (I)i mode, (II)si, (III)s; (c) E2: (I) i mode, (II)si, (III)s, (IV) in phase.

$$H_s = \chi(\rho_p - \rho_n)^2/\rho_0 , \tag{2.4}$$

which arises from the so-called "symmetry energy" term

$$E_s = \chi(N - Z)^2/A; \quad \chi \simeq 20 \text{ MeV} \tag{2.5}$$

in the semiempirical Weizsacker's formula for the masses of nuclei,

$$E = E_{volume} + E_{surface} + E_s + \ldots . \tag{2.6}$$

This more sophisticated theory gives for the frequency of nucleus dipole oscillations

$$\omega^{coll.} \sim \frac{2.08}{R}\sqrt{\frac{8\chi\ NZ}{m^*\ A^2}} \sim \frac{70 \text{ MeV}}{A^{1/2}} , \tag{2.7}$$

(m^* is an effective nucleon mass in nuclear matter), which can be compared with the result of the "hard sphere" model (2.3).

The question arises: can hadronic matter composed of quarks (and apparently, of gluons) develop experimentally observable collective oscillations, as nuclear matter does.

The main point of the paper[1] was that the collective vibration of the hadronic cluster produced via e^+e^--annihilation could explain a rather big value as well as large-scale irregularities in the energy dependence of the total annihilation cross section. We shall study here along this line the hydrodynamic picture of collective resonance phenomena in multiquark hadronic systems.

Consider the case when the numbers of quarks and antiquarks produced in high energy e^+e^--annihilation are large enough so that the qualitative two-fluid description can be applied to the corresponding hadronic system (with the quantum numbers of the photon)(Fig. 4a). We rely on the idea that for large excitations the hadronic vacuum can be described qualitatively as a polarizable classical medium which will resonate on the frequencies determined by collective vibrations of quark-antiquark fluids. To illustrate the idea we consider here the simplest version of the model of hydrodynamic oscillations of two spinless incompressible liquids--that of quarks and antiquarks placed in a spherical volume of a radius R. That is just the analogy of the Goldhaber-Teller "hard sphere" model for the dipole proton-neutron oscillations in spherical nuclei. The model is based on the assumption that the restoring force, preventing the separation of the centers of masses of the two liquids, is proportional to the number of separated particles (shaded regions of Fig. 4b).

For small oscillations the potential of the restoring force is quadratic in ξ, so the Hamiltonian describing these

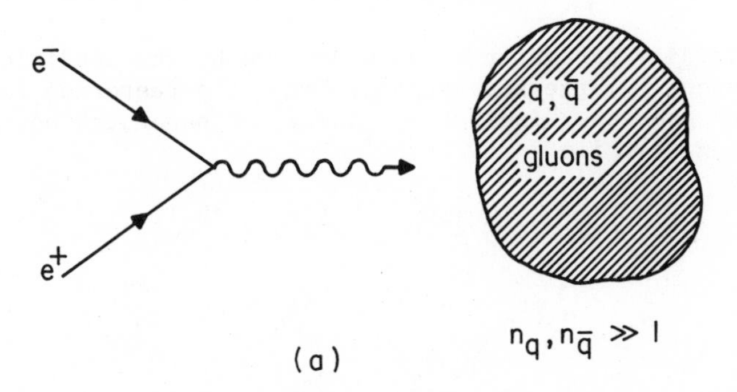

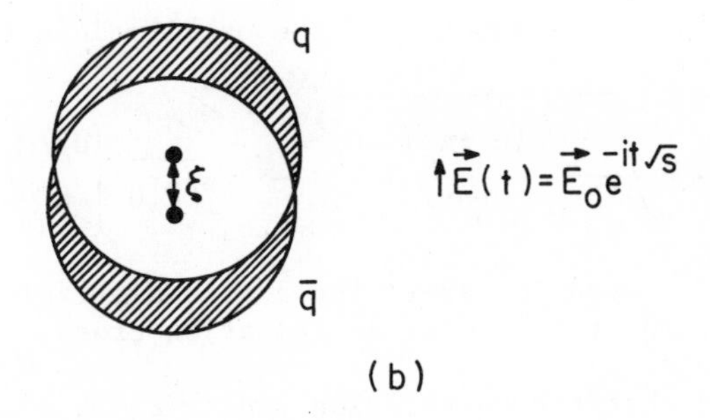

FIGURE 4. (a) Schematic representation of hadron matter production process in one-photon e^+e^--annihilation.
(b) Illustration of the "hard-sphere" model: quark-antiquark splitting in uniform electric field of the virtual photon in the center of mass system.

oscillations has the form

$$H = \frac{1}{2} M_q \frac{n_q n_{\bar{q}}}{n_q + n_{\bar{q}}} \xi^2 + \frac{1}{2} \alpha\xi^2 . \tag{2.8}$$

The rigidity parameter α is determined by the condition that at sufficiently large separations, $\xi = r_0$, the restoring force potential exceeds the "ionization" energy of separated particles (see Fig. 5), i.e.

$$\alpha \frac{1}{2} r_0^2 = U_0(\delta n_q + \delta n_{\bar{q}}), \tag{2.9}$$

where

$$\delta n_q = n_q \frac{\delta V}{V}, \quad \delta n_{\bar{q}} = n_{\bar{q}} \cdot \frac{\delta V}{V},$$
$$\delta V \simeq \pi R^2 r_0 \quad \text{for } r_0 << R, \tag{2.10}$$

and U_0 is an "ionization" energy per one quark.

Thus one has

$$\alpha = 3/2Rr_0 \cdot (n_q + n_{\bar{q}})U_0, \tag{2.11}$$

which gives for the frequency of the oscillations the following result:

$$\omega = \sqrt{\frac{\alpha}{\mu}} = \sqrt{\frac{3}{2Rr_0} \frac{(n_q + n_{\bar{q}})^2}{n_q n_{\bar{q}}} \frac{U_0}{M}} \longrightarrow \sqrt{6/Rr_0} \left(\begin{matrix} U_0/M_q \to 1 \\ n_q = n_{\bar{q}} \end{matrix} \right). \tag{2.12}$$

This result was used in paper[1] for a possible explanation of the energy structure of the total annihilation cross section $\sigma_{e^+e^-} \to$ hadrons.

Assuming that the correlation length r_0 is determined by the mass of quanta mediating the interaction between quarks (gluon?), i.e.

$$1/r_0 \sim m_g , \tag{2.13}$$

and that the size of a hadronic system developing collective oscillations can be underestimated by the inverse total resonance width,

$$R \gtrsim \Gamma_{tot}^{-1}, \tag{2.14}$$

the authors of paper[1] derived the relation[10]

$$\omega_{coll}^2 \lesssim 6 m_g \Gamma_{tot} . \tag{2.15}$$

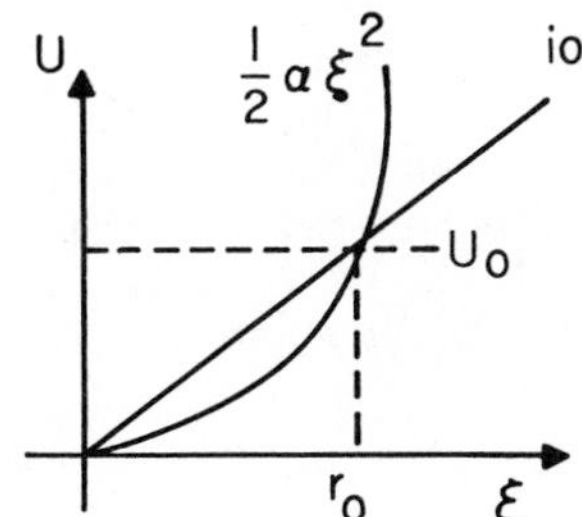

FIGURE 5. Graphic demonstration of the energy balance of two potentials: one of the restoring force $U = \frac{1}{2}\alpha r^2$ and another of the quark ionization of hadronic matter $U_{ioniz.} = U_0(\delta n_q + \delta n_{\bar{q}}) \simeq U_0 \rho \delta V$.

This says that for ω_{coll} and m_g of the order of a few GeV/c^2 the total resonance width should be of the order of a few hundreds MeV or even more. Of course, this model gives only a crude picture of the hadronic collective resonance. It ignores, for instance, a discussion of confinement properties of multiquark systems in the normal hadronic phase.

III. HYDRODYNAMIC OSCILLATIONS IN CONFINED HADRONIC MATTER

Here we shall consider the hydrodynamic oscillations in hadronic matter confined to a finite volume. Due to the assumption that masses of free (unconfined) quarks are very large, we shall describe the hadronic matter by the non-relativistic hydrodynamic equations for a two-component fluid, with ρ_q and $\rho_{\bar{q}}$ being the local densities of quark and antiquark components. We start from the Hamiltonian

$$H = T + U, \tag{3.1}$$

$$T = \frac{1}{2} M_q \int dr(\rho_q V_q^2 + \rho_{\bar{q}} V_{\bar{q}}^2), \quad U = \int dr \mathcal{H}, \tag{3.2}$$

where the potential energy is assumed for simplicity to be local and bilinear in the densities of the quark/antiquark fluids:

$$\mathcal{H} = M_q(\rho_q + \rho_{\bar{q}}) + A(\rho_q^2 + \rho_{\bar{q}}^2) + 2B\rho_q\rho_{\bar{q}} \ . \tag{3.3}$$

By using

$$A = \chi - g, \quad B = -\chi - g < 0 \quad \text{(attraction)}, \tag{3.4}$$

we rewrite Eq. (3.3) as

$$\mathcal{H} = (\rho_q + \rho_{\bar{q}})(M_q - U) + \chi(\rho_q - \rho_{\bar{q}})^2, \tag{3.5}$$

where $U = g(\rho_q + \rho_{\bar{q}})$ is the confinement potential, and the last term in (3.5) is the analogy of the "symmetry energy" potential in the nuclear theory.

First of all we discuss the confinement properties of the hadronic systems described by the Hamiltonian (3.5). Consider first the case

$$\rho_q = \rho_{\bar{q}} \text{ ("symmetric phase").} \tag{3.6}$$

Obviously, the potential energy

$$\mathcal{H} = \rho(M_q - g\rho); \quad \rho = \rho_q + \rho_{\bar{q}} \ , \tag{3.7}$$

has a finite limit for superheavy quarks, as $M_q \to \infty$, for either

$$\rho = 0 \text{ (free quark-antiquark vacuum)} \tag{3.8}$$

or

$$\rho = \rho_0 = \left(\frac{g}{M_q}\right)^{-1} \text{ (confined hadronic phase).} \tag{3.9}$$

Let now

$$\rho_q \neq \rho_{\bar{q}} \text{ ("non-symmetric phase").} \tag{3.10}$$

Consider, for instance, a single quark state. Due to Eq. (3.5) the energy of the state is equal to

$$E(\text{single quark}) = M_q + \frac{\chi - g}{V}, \tag{3.11}$$

and hence one has to put $g=\chi$ to avoid the dependence on the space volume V (no confinement for a single quark).

In general, the states in the symmetric and the non-symmetric phases are split in energies by the amount

$$M_q \cdot (n_q - n_{\bar{q}})^2 / (n_q + n_{\bar{q}}) = \Delta E \equiv E(\text{symmetric}) - E(\text{non-symmetric}). \tag{3.12}$$

To the different kinds of phase transitions in hadronic matter there exist the different kinds of possible collective excitations. In our scheme one can discuss the following types of phase transitions:

A. Symmetric phase $\longleftrightarrow$ Non-symmetric phase, $(\rho_q = \rho_{\bar{q}})$ $\quad$ $(\rho_q \neq \rho_{\bar{q}})$

B. Confined phase $\longleftrightarrow$ Nonconfined phase . $(\chi\rho = M_q)$ $\quad$ $(\chi\rho \neq M_q)$

The corresponding collective excitations could be related with the hydrodynamical fluctuations of the $(\rho_q - \rho_{\bar{q}})$ or $(\rho_q + \rho_{\bar{q}})$ local densities, respectively, which should be considered in that case as compressible fluids. In terms of compressibility the situation can be summarized as follows:

	Compressible	Incompressible
(i)	$\rho_q \pm \rho_{\bar{q}}$	ρ_q , $\rho_{\bar{q}}$ separately
(ii)	$\rho_q - \rho_{\bar{q}}$	$\rho_q + \rho_{\bar{q}}$
(iii)	$\rho_q + \rho_{\bar{q}}$	$\rho_q - \rho_{\bar{q}}$

The first case (i) is just the "hard sphere" model developed by M. Goldhaber and E. Teller. The case (ii) corresponds to the fluctuations around the symmetric phase, and its nuclear analogy is the most successful in describing the giant resonance phenomena in photonuclear reactions. The case (iii) corresponds to the fluctuations around the confined hadronic phase which is characterized by the definite value of the total local density of quarks and antiquarks $\rho_0=M_q/g$. This type of fluctuation raises the problem of stability of the confined hadronic phase and requires a special consideration.

Now we shall only briefly discuss this problem which is, up to our present understanding, beyond the qualitative approach being described here. Consider the symmetric phase $\rho_q = \rho_{\bar{q}}$, so that

$$H = g\rho(\rho_0 - \rho); \quad \rho = \rho_q + \rho_{\bar{q}} . \tag{3.13}$$

The local fluctuations of the total quark/antiquark density around two distinguished levels $\rho=0$ and ρ_0 (see Fig. 6) are characterized by the sign and the magnitude of the derivative

$$\frac{\delta H}{\delta \rho} = \begin{cases} M_q > 0; & \text{at } \rho = 0 \\ -M_q < 0; & \text{at } \rho = \rho_0. \end{cases} \tag{3.14}$$

So, it is unlikely that there could be any collective oscillations around the zero level of the total local density (stability of the "frozen out" quark-antiquark matter or the vacuum). On the other hand, the fluctuations around the confined phase with non-zero local density ρ_0 seem to be very probable. In fact, as we can show, these fluctuations are described under some conditions by an elliptic type partial differential equation, which leads to either imaginary frequencies or imaginary wave lengths. Thus, there are in general exponentially increasing in spacetime as well as exponentially decreasing ones. Those unwanted exponentially growing solutions which are essentially unstable, can be excluded by appropriate boundary conditions.

In what follows we shall consider only the collective oscillations around the confined symmetric phase $\rho_q=\rho_{\bar{q}}=\rho_0/2$ which corresponds to fluctuations of the relative quark-antiquark local density $(\rho_q-\rho_{\bar{q}})$, i.e. the case (ii). The relevant Hamiltonian is given by:

$$H = m_{eff}\rho_0 + \chi(\rho_q - \rho_{\bar{q}})^2 + \frac{1}{2\rho_0} M_q \rho_q \rho_{\bar{q}} V^2, \tag{3.15}$$

where $V=(V_q-V_{\bar{q}})$ is a local field of the quark-antiquark velocity, and $m_{eff} = M_q - g\rho_0$ - the effective mass of confined quarks - is

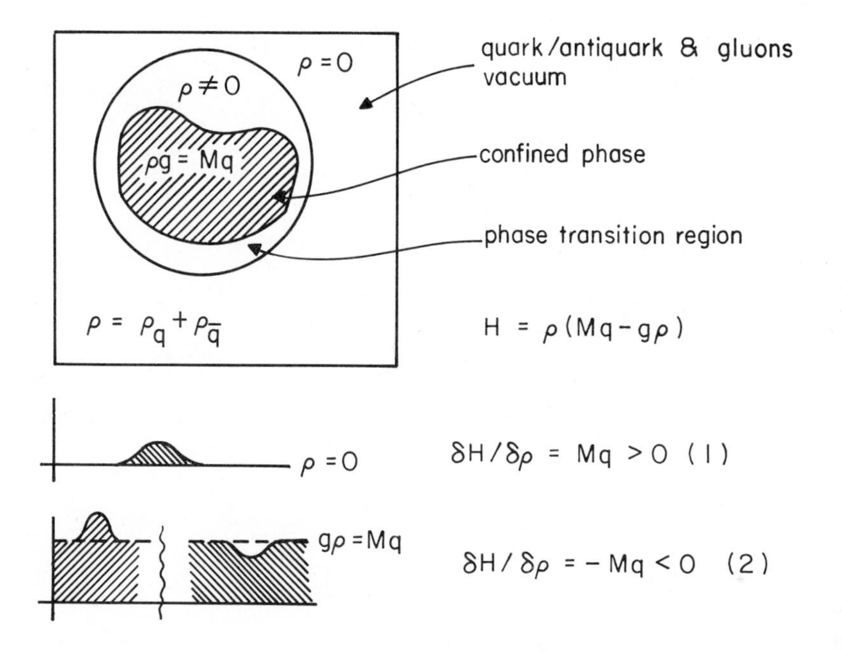

FIGURE 6. Pictures illustrating the different phases of hadronic matter and different types of fluctuations: (1) around quark/antiquark and gluon vacuum, (2) around confined hadronic phase. The sign of the derivative $\delta H/\delta\rho$ characterizes whether the fluctuation is stable or not.

assumed to be negligible ($\simeq 0$). For small oscillations inside rigid spherical surfaces this problem was solved a long time ago.[11] The motion equations take the form

$$M_q \frac{dV}{dt} = \frac{1}{\rho_q \rho_{\bar{q}}} \rho_0 F \simeq -4\chi \,\mathrm{grad}(\rho_q - \rho_{\bar{q}}) \tag{3.16}$$

and assuming that for small hydrodynamic oscillations the velocity flow is potential, i.e.

$$V = -\mathrm{grad}\, \phi, \tag{3.17}$$

one gets

$$\dot{\phi} = \frac{\partial \phi}{\partial t} \simeq \frac{4\chi}{M_q} (\rho_q - \rho_{\bar{q}}) . \tag{3.18}$$

Using Eqs. (3.17) and (3.18) it is easy to rewrite the Hamiltonian in the form

$$H = \frac{1}{8} M_q \rho_0 \left\{ (\mathrm{grad}\, \phi)^2 + \frac{M_q}{2\chi\, \rho_0} \dot{\phi}^2 \right\}, \tag{3.19}$$

which immediately leads to harmonic oscillations with frequencies

$$\omega = k \left(\frac{2\chi\rho}{M_q} \right)^{1/2} . \tag{3.20}$$

The values of the wave vector k are determined from the boundary condition (at r=a)

$$\begin{aligned} V_n &= 0 \quad \text{or} \quad (r \cdot \mathrm{grad}\, \phi) = 0, \quad \ell \geq 1 \\ V_t &= 0 \quad \text{or} \quad \phi = 0 \qquad\qquad\quad , \quad \ell = 0, \end{aligned} \tag{3.21}$$

where ℓ is the angular momentum. The solutions of the problem are of the form

$$\phi \propto Y_\ell^m(\cos\theta, \phi) j_\ell(kr) , \tag{3.22}$$

so the eigenvalues of the wave number $(ka)_\ell$ are determined by the roots of the Bessel functions (for $\ell=0$) or its first derivatives (for $\ell \geq 1$). We list for convenience the first few solutions for the wave number $(ka)_\ell$[12] (see Fig. 7):

The functions $J_n(z), Y_n(z)$ for $n = 0,1,2$

$$J_o(z) = \frac{\sin z}{z}$$

$$J(z) = \frac{\sin z}{z^2} - \frac{\cos z}{z}$$

$$J_2(z) = \frac{3}{z^2} - \frac{1}{z} \quad \sin z - \frac{3}{z^2} \cos z$$

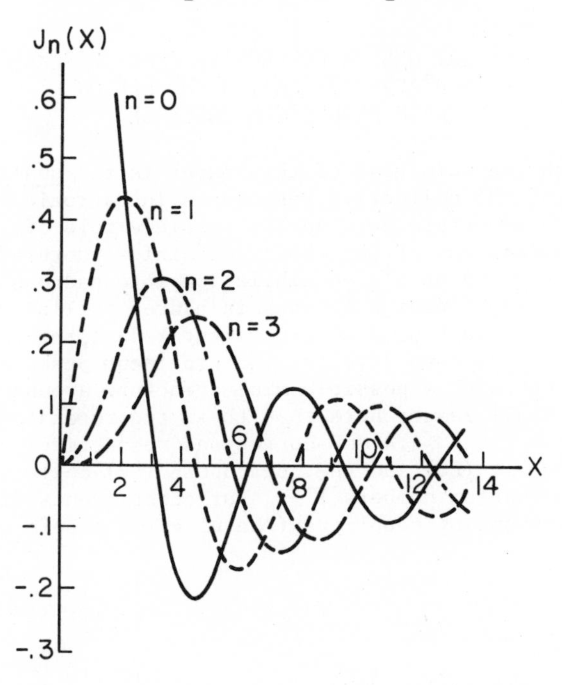

FIGURE 7. The analytic expressions and the graphs of the first few Bessel functions for n=e=0,1 and 2.[11]

		$n_r = 1$	$n_r = 2$
$j_\ell(ka) = 0$	$\ell = 1$	2.08	5.95
	$\ell = 2$	3.31	7.30
	....	....	....
$j_{\ell=0}(ka)=0$	$\ell = 0$	3.14	6.25

IV. SPECULATIONS ON COLLECTIVE QUARK RESONANCES IN e^+e^--ANNIHILATION AND LEPTON PAIR PRODUCTION PROCESSES

Although the main goal of this paper is the qualitative description of the collective resonances in hadronic matter, we would like to speculate here on the possible role of this phenomena in explanation of the observed "mini-structure" of the total cross section of e^+e^--annihilation into hadrons[4] as well as of the excess of prompt leptons in hadron collisions.[3]

As it is known from SLAC experimental data, the total cross section $\sigma_{e^+e^-} \to$ hadrons develops two prominent peaks at about 4.1 and 4.4 GeV with a possible mini-structure around these regions.[4] We shall assume here that these two resonance regions centered at $\sim$4.1 and $\sim$4.4 GeV correspond respectively to the electric monopole (E0) and the quadrupole (E2) collective resonances with quantum numbers 1^{--}. For constituents with spin 1/2 the corresponding couplings (the electric dipole operator) are respectively

$$\text{E0:} \qquad \sim \vec{\sigma} \qquad \left({}^3S_1\right),$$

$$\text{E2:} \qquad \sim \vec{r}(\vec{r} \cdot \vec{\sigma}) - \vec{\sigma}\,\frac{r^2}{3} \qquad \left({}^3D_1\right).$$

We should notice, however, that for quarks with spin 1/2 the dipole oscillations $({}^1P_1)$ are of pure magnetic type (M1) with quantum numbers 1^{++} and hence cannot be observed in e^+e^--annihilation or lepton pair production experiments. On the other hand, if there is any necessity of boson-like constituents in the description of hadronic vacuum fluctuations, the electric dipole collective resonances coupled to a photon via

$$\text{E1:} \quad \sim \vec{r} \quad \left({}^1P_1\right)$$

could be seen in these experiments. Returning to the discussion of the observed peaks in the total cross section of e^+e^--annihilations, we note that the difference in peak energies, which by an assumption is related to the difference in the angular momenta, could give an estimation of the mass of heavy quarks. In fact, from the relation

$$E_{\ell=2} - E_{\ell=0} \simeq \left.\frac{\ell(\ell+1)}{2M_q a^2}\right|_{\ell=0}^{\ell=2} \simeq 300 \text{ MeV}, \tag{4.1}$$

one finds for $1/a \sim 1$ GeV that $M_q \sim 10$ GeV. The ratio of the energies,

$$E_{\ell=2}/E_{\ell=0} \simeq 4.4/4.1 \simeq 1.07, \tag{4.2}$$

is very close to the ratio of the frequencies of quadrupole and monopole collective oscillations listed in the previous section

$$(ka)_{\ell=2}/(ka)_{\ell=0} \simeq 3.31/3.14 \simeq 1.05. \tag{4.3}$$

In absolute scale the required energies of the observed peaks correspond to the following values of the inverse radius of hadronic clusters:

$$\begin{aligned} \frac{1}{a} &\simeq 0.94 \text{ GeV}(\ell=2), \\ \frac{1}{a} &\simeq 0.92 \text{ GeV}(\ell=0). \end{aligned} \tag{4.4}$$

By the way, the energy of dipole oscilaltions can be estimated from the ratio of frequencies

$$\frac{\omega_{\ell=0} + \omega_{\ell=2}}{\omega_{\ell=1}} \simeq 3.1 \tag{4.5}$$

that gives $E_{\ell=1} \simeq 2.74$ GeV or in terms of the inverse value of the radius, $1/a \approx 0.94$ GeV.

Besides the angular momentum splitting in energy one should expect also the appearance of some resonance sub-structure in accordance with isotopic spin and hypercharge assignments of collective resonances. This "mini-structure" is seen, apparently, in the recent SLAC experiments at energies around and above 4 GeV. Obviously, the "hidden" charm gives the highest jump in energy, say

$$\Delta E^2 \approx m_\psi^2 - m_\omega^2 \simeq 8.97 \text{ GeV}^2, \tag{4.6}$$

leading to the new peaks ("charm-mirror") with the energies[13]:

$$4.1 \text{ GeV} \to 5.08 \text{ GeV},$$
$$4.4 \text{ GeV} \to 5.32 \text{ GeV}, \tag{4.7}$$

which we will denote by $E_\psi(E0)$ and $E_\psi(E2)$.

The next question is what can one say about widths of collective resonance? The experience of the theory of nuclear giant resonance phenomena teaches us that this problem is the hardest one for the qualitative hydrodynamical description. Evidently, the total widths of collective resonances must exceed the average energy of excitation of separate single quark degrees of freedom in a region with a finite size a:

$$\Delta E_q \sim \frac{1}{2M_q} \frac{1}{a^2} . \tag{4.8}$$

Thus, one has

$$\Gamma_{tot} > \Delta E_q \sim \alpha\omega^2/M_q, \tag{4.9}$$

where

$$\begin{aligned} \alpha &= 2.28 \times 10^{-2} \ (\ell=2), \\ &= 2.54 \times 10^{-2} \ (\ell=0), \\ &= 5.78 \times 10^{-2} \ (\ell=1). \end{aligned} \tag{4.10}$$

So, for example, for the values

$$\frac{1}{a} \sim 1 \text{ GeV} \quad \text{and} \quad M_q \sim 10 \text{ GeV}$$

one gets for the lower bound of total resonance widths the value 50 MeV.[14] More general situations are shown in Fig. 8.

It is worth noting that there could exist, in general, collective oscillations in hadronic matter with non-zero color and frequencies comparable to those of uncolored ones. We emphasize however, that colored collective resonances should be stable with respect to strong decays because of the color conservation and can be observed (if any exist) only through their leptonic decays (obviously, under the condition that the electromagnetic or weak currents have colored counterparts[15]). Thus the previous estimation on resonance widths does not work for colored collective excitations in hadronic matter.

V. "CHROMO-HYDRODYNAMICS"

In the QCD-Bag theory,[16] the most popular approach to quark confinement at the present time, there are colored vector gluons

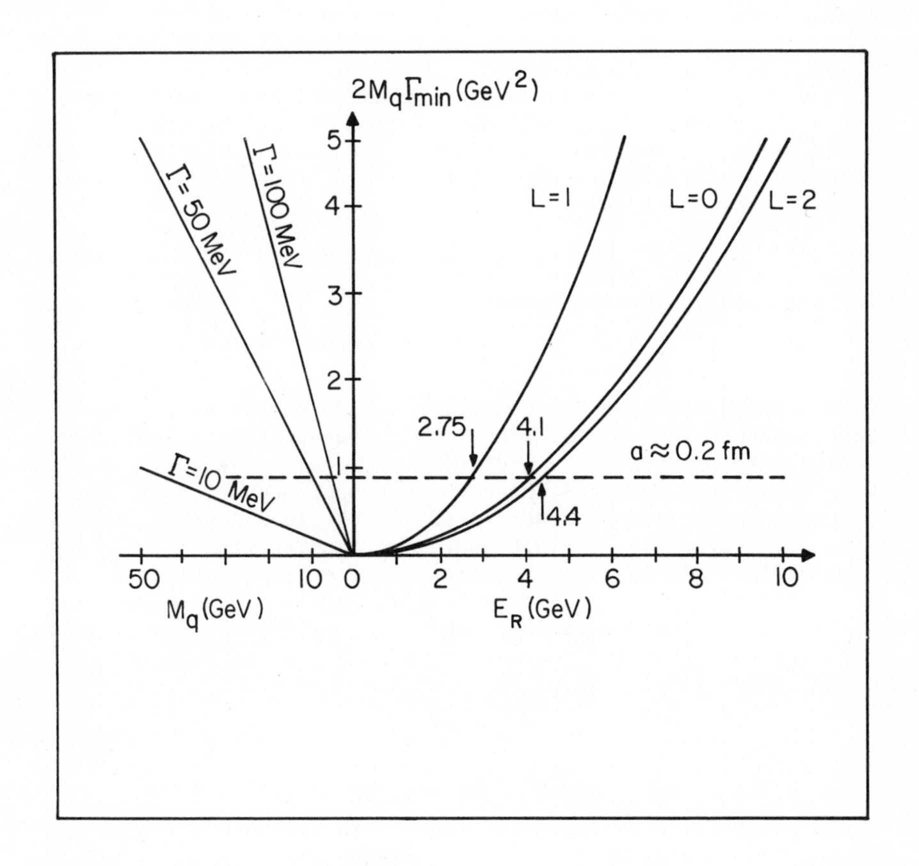

FIGURE 8. The illustration of the relations between the energies of the collective resonances for modes E0(L=0), E1(L=1) and E2(L=2), the unconfined quark mass M_q and the lower bound on the total resonance widths Γ_{min}. The arrows indicate the positions of the collective resonances for the particular value of a size of hadronic system a $\approx$ 0.2 fm.

confined to the same interior as quark and antiquark constituents. Thus, there arises the problem of incorporating gluon degrees of freedom into our scheme.

Here we shall not introduce a condensed phase for gluons, but shall assume a classical description of a gluon field interacting with quark-antiquark matter. Apparently, this approach, which we shall call hereafter "chromo-hydrodynamics", could be based on the equation:

$$\partial_\nu T_{\mu\nu} = 0 \text{ (inside a finite volume V)}, \tag{5.1}$$

for the energy-momentum tensor

$$T_{\mu\nu} = T_{\mu\nu}^{\text{quark fluids}} + T_{\mu\nu}^{\text{gluon fields}}, \tag{5.2}$$

with appropriate boundary conditions which ensure confinement. Unfortunately there is no theory of a non-Abelian field interacting with a macroscopic medium.[17] For this reason, we consider here as a first step the chromo-hydrodynamic equations in the lowest (non-vanishing) order in the quark-gluon coupling constant g. Obviously, in this approximation the gluon fields become effectively Abelian ones like the electromagnetic field. The underlying physics is very simple. Assume that there are a number of fluctuation regions, where $(\rho_q - \rho_{\bar{q}}) \neq 0$; the fluctuations generate the gluon fields, say E^a and B^a, satisfying the Maxwell equations, e.g.,

$$\begin{aligned} \text{div } E^a &= 4\pi g^a (\rho_q - \rho_{\bar{q}}) \\ \text{div } B^a &= 0, \text{ etc. }, \end{aligned} \tag{5.3}$$

where a=1,2,...,8 is the color index. In turn these fields influence the motion of quarks and antiquarks and hence on developing of the fluid fluctuations. The relevant notion describing this influence is the so-called "ponderomotive" or Lorentz force[18]:

$$\vec{F} = \sum_a g^a (\vec{E}^a + \frac{1}{c} \vec{V} \times \vec{B}^a). \tag{5.4}$$

In terms of the action principle we have

$$\delta A = \delta \int dt(T+U) + \int dr\, \vec{F}(\rho_q \delta\vec{\xi}_q - \rho_{\bar{q}} \delta\vec{\xi}_{\bar{q}}) = 0, \tag{5.5}$$

where $\delta\xi_q$ and $\delta\xi_{\bar{q}}$ are the fields of displacements of quark and antiquark positions, and

$$T = \frac{1}{2} M_q \int dr \left[\frac{\rho_q \rho_{\bar{q}}}{\rho_0} v^2 + \rho_0 V^2 \right], \qquad (5.6)$$

$$U = \chi \int dr(\rho_q - \rho_{\bar{q}})^2.$$

Here we use the notations

$$\delta\vec{\xi}_q = \delta\vec{R} + \delta\vec{\xi} \cdot \rho_{\bar{q}}/\rho_0 ,$$
$$\delta\vec{\xi}_{\bar{q}} = \delta\vec{R} - \delta\vec{\xi} \cdot \rho_q/\rho_0 , \qquad (5.7)$$

where $\delta\xi=(\delta\xi_q-\delta\xi_{\bar{q}})$, $\rho_0= (\rho_q+\rho_{\bar{q}})$ = const, so that $v = \delta\xi/\delta t$; $V = \delta R/\delta t$. From the continuity equations it follows that

$$\frac{d\rho}{dt} = \frac{\partial\rho}{\partial t} + (v\ \mathrm{grad})\rho = -\rho\ \mathrm{div}\ V, \qquad (5.8)$$

for q and $\bar{q}$, and hence we can put in the action principle

$$\delta\rho_q = -\rho_q \cdot \mathrm{div}\ \delta\xi_q,$$
$$\delta\rho_{\bar{q}} = -\rho_{\bar{q}} \cdot \mathrm{div}\ \delta\xi_{\bar{q}}, \qquad (5.9)$$

so that

$$\delta U = \chi \int dr(\delta\vec{R} \cdot \mathrm{grad})(\rho_q-\rho_{\bar{q}})^2 +$$
$$\frac{4\chi}{\rho_0} \int dr\ \rho_q\rho_{\bar{q}}(\delta\vec{\xi} \cdot \mathrm{grad})(\rho_q-\rho_{\bar{q}}) . \qquad (5.10)$$

Now one obtains the following motion equation for the velocity of relative oscilaltions of the quark and antiquark fluids:

$$\frac{d\vec{V}}{dt} = - \frac{4\chi}{M_q} \mathrm{grad}(\rho_q - \rho_{\bar{q}}) + \frac{2}{M_q} \cdot \vec{F}. \qquad (5.11)$$

The boundary condition for a spherical shape of the volume V occupied by $q/\bar{q}$ fluids is

$$(\vec{r} \cdot \vec{v})_{r=a} = 0, \qquad (5.12)$$

so no effects of surface vibrations are considered.

By applying the divergence operation to both sides of Eq. (5.11) and using the relation

$$\mathrm{div}\ V=\mathrm{div}(v_q-v_{\bar{q}})= - \frac{1}{\rho_q} \frac{d\rho_q}{dt} + \frac{1}{\rho_{\bar{q}}} \frac{d\rho_{\bar{q}}}{dt} = - \frac{1}{2} \frac{\rho_0}{\rho_q\rho_{\bar{q}}} \frac{d}{dt}(\rho_q-\rho_{\bar{q}}) \qquad (5.13)$$

one gets the equation

$$\left[\frac{d}{dt}\frac{1}{c^2}\frac{d}{dt} - \Delta + \frac{2\pi}{M_q} g_a^2\right](\rho_q - \rho_{\bar{q}}) = 0, \tag{5.14a}$$

or

$$\left[\frac{1}{c^2}\frac{d^2}{dt^2} + \frac{16\chi}{M_q\rho_0}\left(\frac{1}{c^2}\frac{d}{dt}\rho_q\right)^2 - \Delta + \frac{2\pi}{M_q} g_a^2\right](\rho_q - \rho_{\bar{q}}) = 0, \tag{5.14b}$$

where $c^2 = 8\chi\, \rho_q\rho_{\bar{q}}/M_q\rho_0$.

We shall assume for small oscillations that

$$\frac{d\rho}{dt} \approx \frac{\partial\rho}{\partial t} \quad \text{or} \quad (v\ \text{grad})\rho << \frac{\partial\rho}{\partial t}, \tag{5.15}$$

which means

$$\tau V_q << \ell, \tag{5.16}$$

where τV_q is an average path of particles during one period of oscillations $\tau \sim 1/\omega$, and ℓ a length on which $\rho_{q/\bar{q}}$ varies more or less considerably.

The resulting equation for linearized hydrodynamical oscillations of the quark and antiquark fluids interacting with gluon fields takes the form of the wave equation, i.e.

$$\left(\Delta - \frac{1}{c^2}\frac{\partial^2}{\partial t^2} - c^2\mu^2\right)\delta\rho = 0; \quad \delta\rho = (\rho_q - \rho_{\bar{q}}). \tag{5.17}$$

Here $c = (8\chi\, \rho_q^{(0)}\rho_{\bar{q}}^{(0)}/M_q\rho_0)^{\frac{1}{2}}$ is an effective velocity and $\mu = (\pi g_a^2/\chi)^{\frac{1}{2}}$ an effective mass of quanta of the collective excitations. For the frequency of the collective oscillations we get

$$\omega = \left(\frac{8\chi}{M_q}\frac{\rho_q\rho_{\bar{q}}}{\rho_0}\right)^{1/2} \cdot \left(k^2 + \frac{2\pi g_a^2}{M_q}\right)^{1/2} \tag{5.18}$$

where the values of k^2 are determined by the boundary condition

$$0 = (r\cdot\text{grad})\delta\rho \left|\begin{array}{l} r=0 \\ \\ r=a \end{array}\right. \begin{array}{c} \\ \rightarrow \\ \\ \end{array} \begin{array}{ll} j_e'(ka) = 0 & e \geqslant 1, \\ \\ j_0(ka) = 0 & e = 0. \end{array} \tag{5.19}$$

By using $\chi\rho_0/M_q \rightarrow 1$ and $\rho_q = \rho_{\bar{q}} = \rho_0/2$ as it should be for the confined hadronic phase we get finally

$$\omega = \sqrt{2k^2 + \omega_p^2}, \tag{5.20}$$

where

$$\omega_p^2 = \frac{4\pi g_a^2}{M_q} \rho_0 . \tag{5.21}$$

It is interesting to note that the equation determining ω_p is the well-known formula for the frequency of collective oscillations in plasma[19] (here the "plasma" made of the massive quarks and antiquarks interacting through the gluon fields!). This amusing analogy brings an idea of the possible existence of a new state of hadronic matter, namely the gas of free (unconfined) quarks and antiquarks interacting with the colored gluon fields imbedded in the confined hadronic phase. Obviously, no such "plasma" type state can be realized in the vacuum.

Now we estimate the value of the hadronic "plasma" frequency ω_p. First of all, we should take into account the color group structure of the quark-gluon coupling constant. We guess that this can be done by substitution

$$g_a^2 \to \frac{1}{3} \text{ Spur } (g_* \lambda_a)^2 = g_*^2 C_2^{\text{color}},$$

where C_2^{color} is the quadratic Casimir operator for the SU_3'-color group, so that

$$C_2^{\text{color}}(\text{single quark}) = 16/3.$$

Here g_*--the rationalized coupling constant--is to be taken in Gauss units, so that

$$g_*^2 = g^2/4\pi$$

where g is the usual quark-gluon coupling constant determined by the Lagrangian $g\bar{\psi}\lambda^a\gamma_\mu\psi A_\mu{}^a$. One has now

$$\omega_p^2 = \frac{g^2\rho_0}{M_q} C_2^{\text{color}} = \left(\frac{g^2}{4\pi}\right)\frac{16}{M_q a^3} .$$

This can be compared with the energy of self-interaction of a quark through the gluon field:

$$\Delta M(\text{self-inter.}) = \left(\frac{g^2}{4\pi}\right) \frac{C_2^{\text{color}}}{a} = \left(\frac{g^2}{4\pi}\right) \frac{16}{3a} ,$$

which for $1/a \approx 0.94$ GeV and $g^2/4\pi \sim 1$ gives $\Delta M \sim 5$ GeV. By using Eq. (4.9) one gets the relation

$$\omega_p^2 = 6\Gamma\Delta M$$

which gives under the previous conditions the value $\omega_p^2 \sim \Gamma \cdot 30$ GeV or $\omega_p \sim 1.2$ GeV for $\Gamma \sim 50$ MeV.

Note Added. During the completion of this work, a paper by J.W. Moffat (University of Toronto preprint) was called to my attention in which a "quark-nucleus" shell model of hadrons is proposed on the basis of a generalization of the empirical nucleus mass formula with the symmetry-energy term. This paper does not consider the collective structure of multiquark hadronic states.

REFERENCES

1. V. A. Kuzmin, V. M. Lobashev, V. A. Matveev and A. N. Tavkhalidze, preprint-JINR, E2-8742 (1975), unpublished.

2. M. Goldhaber and E. Teller, Phys. Rev. 74, 1046 (1948).

3. D. C. Hom, L. M. Lederman, et al., Phys. Rev. Lett. 36, 1236 (1976).

4. J. Siegrist, G. S. Abrams, et al., SLAC-PUB-1717, LBL-4804 (1974) (to be published in Phys. Rev. Lett.);B.Richter, SLAC-PUB-1706 (1976).

5. J. P. Davidson, Collective Models of the Nucleus, Academic Press, New York and London, 1968; B. M. Spicer, The Giant Dipole Resonance, Advances in Nuclear Physics, Vol. 2, Plenum Press, New York, 1969.

6. A. E. Glassgold, Warren Hekcrotte, and Kenneth M. Watson, Annals of Phys. 6, 1-36 (1959); H. Überall, Nuovo Cimento XLI, 25 (1966).

7. E. P. Wigner, Phys. Rev. 51, 106 (1937).

8. A. B. Migdal, Journ. of Phys. USSR, 8, 331 (1944).

9. H. Steinwedel and J.H.D. Jensen, Z. Naturforsch, 5a, 413 (1950); H. Steinwedel and P. Jensen, Phys. Rev. 79, 1019 (1950).

10. Obviously, the energies of the collective resonances in the "hard-sphere" oscillation model are related to the frequency by $E(E0)=E(E2)=2E(E1)=2\hbar\omega^{coll}$.

11. Rayleigh, Proc. Lond. Math., Sec. (1) iv.93 (1872) and Theory of Sound, Art. 331; Sir Horace Lamb, Hydrodynamics, Dover Publications, New York.

12. Handbook of Mathematical Functions, National Bureau of Standards, Applied Mathematics Series 55, 1964.

13. We have used the quadratic mass formula $E_{\psi}^{2}(e)-E^{2}(e) \simeq m_{\psi}^{2}-m_{\omega,\rho}^{2}$.

14. We guess that more realistic estimation should include the total quark number, so e.g. for a four quark system $\Gamma_{tot}(2\bar{q}2q) \gtrsim 200$ MeV.

15. A. Tavkhelidze, Proc. Seminar on High Energy Physics and Elementary Particles, Trieste, published by International Atomic Energy, Vienna, (1965), pp.763-779; M. Y. Han and Y. Nambu, Phys. Rev. B139, 1006 (1965).

16. A. Chodos, R. L. Jaffe, K. Johnson, C. B. Thorn, and V. F. Weisskopf, Phys. Rev. D9, 3471 (1974); A. Chodos, R. L. Jaffe, K. Johnson and C. B. Thorn, Phys. Rev. D10, 2599 (1974).

17. The formalism of the non-integrable phase factors developed by C. N. Yang (see, for example, C. N. Yang, "Gauge Fields", the lectures on the 1975 Hawaii Conference, University Press of Hawaii, 1976) could be used to serve as a basis for such a theory.

18. J. L. Synge, Relativity: The Special Theory, North-Holland Publishing Company, Amsterdam, 1965, pp.393-395.

19. James E. Drummond, Plasma Physics, McGraw-Hill Book Co.,Inc., New York-Toronto-London, 1961.

20. R. L. Bramblett, J. T. Caldwell, R. R.Harvey, and S. C. Fultz, Phys. Rev. 133B, 869 (1964).

Is There a Signal of Quark Confinement from Perturbation Theory?

ENRICO C. POGGIO*
Lyman Laboratory of Physics
Harvard University
Cambridge, Massachusetts

and

Physics Department
Brandeis University
Waltham, Massachusetts

It has been widely believed that understanding the infrared structure of non-Abelian gauge theories of colored quarks and gluons (Quantum chromodynamics, Q.C.D.) might bring to light the dynamical mechanism that binds quarks into color singlet states, the hadrons. Unlike all four dimensional, renormalizable Abelian theories, the renormalization group analysis of Q.C.D. is characterized by the fact that the origin in the effective coupling parameter, $g(k^2)$ (i.e., the invariant charge), is unstable at long-distances, $k^2 \to 0$. This is reflected in perturbative calculations of Green's functions in the long distance sector by the appearance of large logarithms. These, a priori, cannot be controlled because one doesn't have a complete knowledge of the fundamental renormalization group function $\beta(g(k^2))$ as $k^2 \to 0$,

$$k^2 \frac{\partial}{\partial k^2} g(k^2) = \beta(g(k^2)) \ . \tag{1}$$

I will return later to comment on this important problem and its implications and relevance to the confinement problem.

Naturally, the first and simplest question that can be asked is whether the presence of the large infrared logarithms affects in any sense the determination of "physical" amplitudes involving quarks and gluons. In the past year or so, various investigations have been performed, all within the framework of perturbation theory, in order to attempt to answer the above

*Work supported in part by the National Science Foundation under Grant No. MPS75-20427.

question. I will now report the results of some of these investigations and I will later attempt to give my global impression of their nature and of what they mean as far as the confinement issue is concerned. To keep a proper perspective I will try to make a comparison with analogous Q.E.D. processes, where the corresponding infrared aspects are completely understood.

I. THE "WIDE-ANGLE" BEHAVIOR OF THE QUARK FORM FACTOR IN THE LEADING LOGARITHM APPROXIMATION

Consider the photon-fermion-anti-fermion vertex function of Fig. 1.

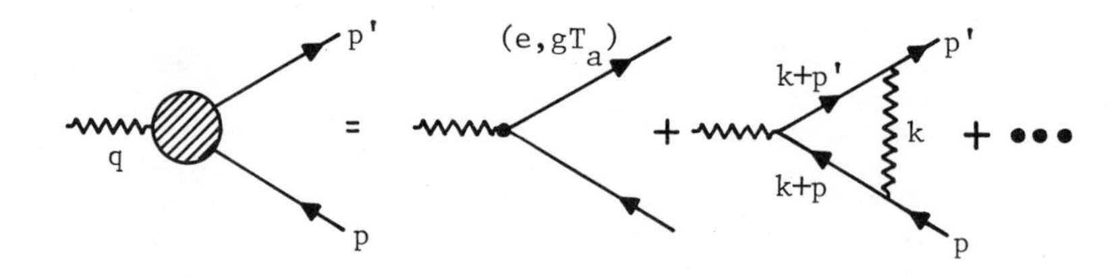

Figure 1

When $q^2 >> m^2 >> (p^2-m^2)$, (p'^2-m^2), we find that the lowest order integral behaves as

$$F^{(1)}(q^2) \sim q^2 \int \frac{d^4k}{(k^2)[k^2+2p\cdot k+(p^2-m^2)][k^2+2p'\cdot k+(p'^2-m^2)]}$$

$$\sim \log \frac{q^2}{m^2} \log \frac{q^2}{p^2-m^2} \, , \qquad (2)$$

that is the integral is logarithmic infrared divergent as the fermions approach their mass shell. In QED it can be shown that: a) only ladder and crossed-ladder diagrams give leading logarithmic contributions, a factor of $(g^2 \log^2 q^2)$ per every exchanged photon, and b) the sum of all contributions exponentiates simply[1]

$$F_{QED}(q^2) = \exp \left| -Ae^2 \log \frac{q^2}{m^2} \log \frac{q^2}{p^2-m^2} \right| , \qquad (3)$$

where A is a positive numerical coefficient obtained from the lowest order calculation. The exponentiation proof follows from a simple eikonal argument which can be carried out because of the simplicity of the structure of QED ladder and crossed-ladder graphs. In QCD all QED-type graphs have the same q^2 behavior as QED but they also have a characteristic group theoretical weight associated with each one of them. The sum of ladder and crossed-ladder graphs alone does not exponentiate. There are, though, other graphs, special to QCD, involving tri-gluon interactions that have also leading logarithmic terms. For example, at the two-loop level, the ladder and crossed ladder contribution to the quark form -factor gives $\Gamma^{(4)}_{QCD}$ (ladder+crossed-ladder) =

$$= \frac{1}{2!}\left\{-AC_N g^2 \log\frac{q^2}{m^2} \log\frac{q^2}{p^2-m^2} + B\, C_N C_A g^4 (\log\frac{q^2}{m^2} \log\frac{q^2}{p^2-m^2}\right\}. \tag{4}$$

A computation of the contribution from the QCD graphs of Fig. 2

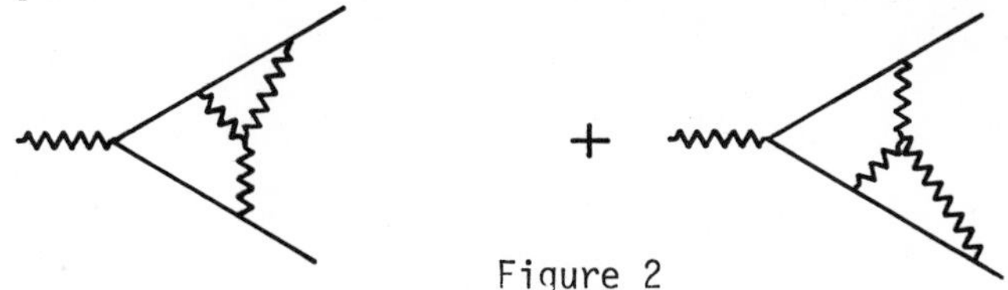

Figure 2

gives the result

$$\Gamma^{(4)}_{QCD}(\text{trigluon}) = -\, BC_N C_A g^4 (\log \frac{q^2}{m^2} \log \frac{q^2}{p^2-m^2}),$$

where $C_N = T^a T_a$ and $if_{abc}\, T_b T_c = -C_A T_a$. To this order, the exponential pattern is seen to be maintained in QCD. Remarkably enough, the computation to the next order shows that by adding all graphs with trigluon vertices, such as that in Fig. 3, to the ladder and crossed ladder ones, after some rather

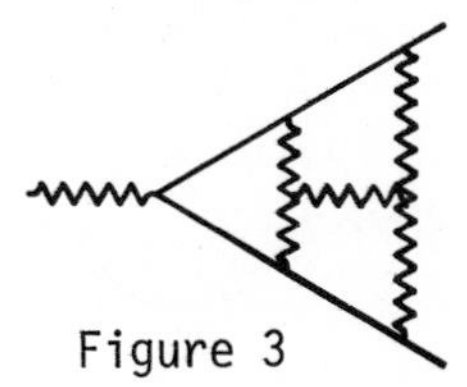

Figure 3

"miraculous" cancellations gives the result[2,3]

$$\Gamma^{(6)}_{QCD} = \frac{1}{3!}\left\{-AC_N g^2 \log^2 q^2\right\} . \tag{5}$$

The QED and QCD result seem to be related by the simple substitution $Ae^2 \leftrightarrow AC_N g^2$. Whether the exponentiation persists in higher order, as far as I know, yet unproven.

II. FIXED-WIDE-ANGLE QUARK-ANTIQUARK SCATTERING

The process in question is depicted in Fig. 4. The region of interest is given by taking $s,t >> m^2 >> (p^2-m^2)$ with s/t fixed. Again, simple combinatoric arguments show that, in QED,

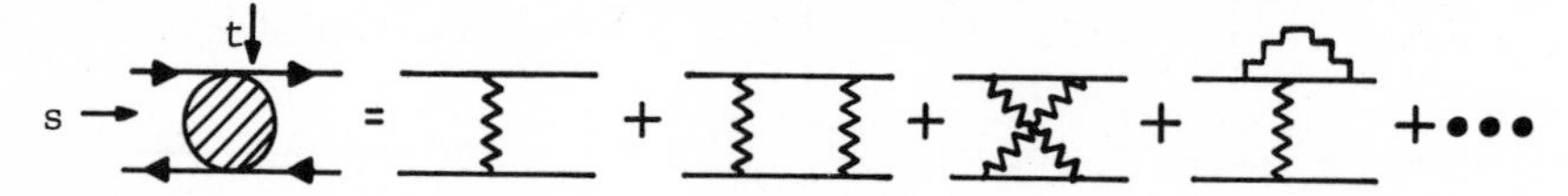

Figure 4

the scattering amplitude in the leading logarithmic approximation behaves as[4]

$$A(s;t) = (\text{Born Term})\ F^2_{QED}(s,e^2), \tag{6}$$

where F is the form-factor of Eq. (3). Again, only after some arduous computations and group theory manipulations resulting in spectacular cancellations, the QCD amplitude to sixth-order is found to obey (6) with the substitution $e^2 \leftrightarrow C_N g^2$. Hopes of generalizing the result are again not apparent.

III. THE THRESHOLD BEHAVIOR OF THE HADRONIC VACUUM POLARIZATION FUNCTION

Let us consider the process $e^+e^- \rightarrow$ hadrons. Of particular interest is the function $R(s) = (12\pi s/\alpha)\sigma_{TOT}(s) = \pi(s+i\varepsilon) - \pi(s-i\varepsilon) = \text{disc}\ \pi(s)$, where $\pi(s)$ is the hadronic vacuum polarization function (see Fig. 5). Let me remind you of the analogous

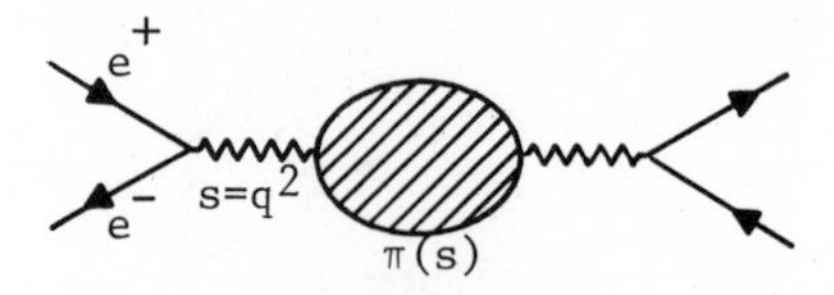

Figure 5

situation in QED. At the onset of an electron-positron threshold, the perturbation theory breaks down. In particular, the contribution to $\pi(s)$ from the diagram with N-ladder photon exchanges (Fig. 6), behaves, near $s=4m^2$, as: (phase-space) $\times$ $(\alpha/v)^N$, where $v = \sqrt{s-4m^2/s}$. Any other Nth order diagram will be

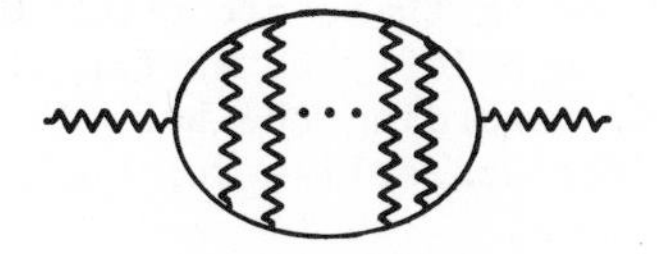

Figure 6

less divergent by factors of the type $v^A \log^B v$. As is well known, the sum to all orders of the most divergent terms yields a sequence of poles at the energies of the positronium bound states. The perturbative analysis in QCD follows similar lines. Analogous diagrams to Fig. 6, with exchanged gluons, have the same $1/v$ behavior. Typical QCD graphs, such as in Fig. 7, give a non-leading behavior. No other dramatic or more severe

Figure 7

singularities are seemingly present so as to indicate that instead of free quarks and gluons one is really dealing with hadrons. Interestingly enough, for the case of heavy "charmed" quarks, because of asymptotic freedom, one can nevertheless use perturbation theory with moderate success to predict the presence of positronium-like states[6] and determine some of their physical features and also understand the general features of "charmed" hadronic data[7] even without fully understanding the details of the binding mechanism. Ironically, it is the absence of some of the "dramatic new singularities" that enabled Appelquist and Politzer[6] to arrive at some of their results. Namely, that for QCD, there is no (perturbative) breakdown of the Kinoshita-Lee-Nauemberg (KLN) theorem.

IV. THE WEAK KLN THEOREM

Basically for QED the KLN theorem[8] states that in computing disc $\pi(s)$ (and therefore R and σ_{TOT}), even though the intermediate states are infrared divergent, the <u>sum over all</u> intermediate states is finite. Furthermore, the cancellation of the divergences occurs <u>graph by graph</u>.

Recently, Helen Quinn[9] and myself discovered that the theorem is in fact valid for any renormalizable theory, in perturbation theory, including therefore QCD. Our proof is simple enough that I would like briefly to sketch it. The result is a corollary of a theorem regarding the behavior of zero-mass Green's functions in the space-like region.

Theorem: Any One-particle Irreducible N-pt Function is Infrared Convergent at Euclidean Non-exceptional Momenta. The proof is done by an iterative procedure and it uses a simple power counting analysis. The iterative scheme consists in proving two steps: A) any renormalized 1PI 2-, 3-,4-point function behaves respectively as

$$\Gamma^{(2)}(k) \sim k^{2-\varepsilon}$$
$$\Gamma^{(3)}(k) \sim k^{1-\varepsilon}$$
$$\Gamma^{(4)}(k) \sim k^{0-\varepsilon} , \qquad (7)$$

as all their external momenta k are scaled to zero simultaneously. B) The theorem is true. The symbol ε is just a notation to remind us that, in perturbation theory, there are logarithmic corrections: the lowest order two-point function behaves as $k^2 \log k^2$. In any arbitrary order of perturbation theory, whatever the corresponding power of $\log k^2$, it will not change the dimensional power behavior of $\Gamma^{(2)}$, $\Gamma^{(3)}$ and $\Gamma^{(4)}$. In the power counting analysis, ε is truly an infinitesimal quantity.

Sketch of the Proof. Notice that A) is true in lowest order. Consider any arbitrary graph contributing to the N-point function. Define its skeletons by shrinking to a point all 2-, 3-, 4-point insertions. Such a concept of skeleton is non-unique, but we are not interested in counting overall weights, we just want to determine the overall power behavior. The skeleton consists of:

N external lines

I internal lines

V_2, V_3, V_4 2-, 3-, 4-point vertices,

so that

$$2I = 2V_2 + 3V_3 + 4V_4 - N,$$

and the number of loops, L, will be

$$L = I - (V_2 + V_3 + V_4) + 1 .$$

When all external momenta vanish simultaneously the degree of divergence of the graph will be given by

$$D_N = 4L-2I + (2-\varepsilon)V_2 + (1-\varepsilon)V_3 + (0-\varepsilon)V_4 = 4 - N - \varepsilon'$$

proving statement A). I should warn you that a rigorous proof of this statement is a bit more involved. Fortunately this has already been done, since A) is nothing but a re-expression of Weinberg's famous power counting theorem. Recall that for massless theories, whether you want to determine large or small momenta behavior is just a simple scaling question. Finally, let all external lines carry some finite momenta. These will flow through the diagram in all the possible ways allowed by energy-momentum conservation. They will therefore serve as an infrared cut-off to all the internal lines they flow through. Define now a reduced graph by shrinking to a point all internal lines carrying the finite external momenta. The reduced graph will have a "super-vertex" formed by the shrinking of the lines, consisting of the N-external lines and m internal lines connecting to a graph $\tilde{G}_R$ (which may be disconnected)(see Fig. 8) which

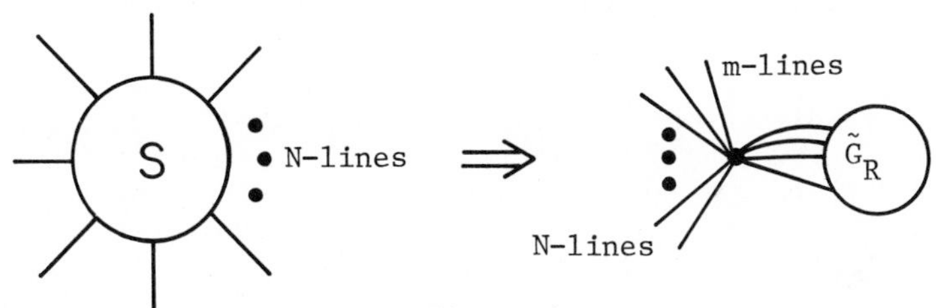

Figure 8

is dependent on the internal momenta only. The overall reduced diagram consists of $\tilde{V}_2$, $\tilde{V}_3$, $\tilde{V}_4$ 2-, 3-, 4-point vertices, $\tilde{I}$ internal lines and $\tilde{L}$ internal loops such that

$$2\tilde{I} = 2\tilde{V}_2 + 3\tilde{V}_3 + 4\tilde{V}_4 + m$$

$$\tilde{L} = \tilde{I} - (\tilde{V}_2 + \tilde{V}_3 + \tilde{V}_4) + 1$$

$$\tilde{D}_N = 4\tilde{L} - 2\tilde{I} + (2-\varepsilon)\tilde{V}_2 + (1-\varepsilon)\tilde{V}_3 + (0-\varepsilon)\tilde{V}_4$$

$$= m - \varepsilon' \quad .$$

Because we are interested in 1PI-graphs, $m>2$. The theorem also holds if some fields are massive. For a proof of this and for a more thorough proof of the above, the interested reader is referred to the original article.

Consider now the vacuum polarization function $\pi(s)$. The coefficients of its perturbation expansion are an analytic function of the external momenta p_i^2 everywhere except where the Landau-Cutkoski rules tell us there are physical singularities. Then $\pi(s)$ is well defined and infrared non-singular except for real time-like s, where it has its physical cuts. In particular we can continue from Euclidean s to evaluate $\pi(s+i\varepsilon)$ and $\pi(s-i\varepsilon)$ for real time-like s and any ε. Therefore (disc $\pi(s)$) is infrared convergent.

Notice that in QED, the cancellation of I.R. divergences is well understood to follow generally a familiar pattern[10]: the infrared singularities due to the exchange of virtual photons are cancelled by singular terms involving the emission of real photons. The validity of the theorem just proved for (disc $\pi(s)$) indicates that, at least for this total amplitude, this is also valid in QCD. I would like to refer to the above result as the "weak" KLN theorem, because in QED and apparently also in QCD, as I will soon discuss, the virtual-real cancellation exists even under more stringent conditions.

V. THE STRONG KLN THEOREM: THE BLOCH-NORDSIECK PROGRAM

The well known Bloch-Nordsieck analysis in QED states that: the infrared divergences associated with virtual corrections are cancelled by corresponding divergences in the emission of undetected photons whose total energy is less than the energy resolution ΔE of the detector. What I will then refer to as the "strong" KLN theorem is the statement that transition amplitudes corresponding to finite-energy detectors are free of infrared divergences. Specifically, the partial cross-section $R_{\Delta E}$ for the inclusive detection of a massive fermion, constructed by summing over all unitarity cuts of $\pi(s)$, restricted by the detector kinematics (ΔE-cuts), will be free of infrared divergences, even though the individual cuts are not. A computation of $R_{\Delta E}$ to lowest non-trivial order containing gluon self-coupling effects has recently been carried out.[11] The details and the physics of this beautiful calculation have been discussed by J. Carazzone in these proceedings. I will limit myself to stating their result: $R_{\Delta E}$ has been verified to be free of IR divergences to the order computed. Again, the finiteness can be determined only after some lengthy computations where some rather remarkable cancellations appear. Furthermore, unlike QED, the systematics of all the cancellations are not yet well understood and the proof of finiteness cannot be extended to all orders.

VI. PERTURBATION THEORY AND CONFINEMENT

Up to now I have presented the results of some calculational "experiments" regarding various aspects of the infrared structure of QCD. I have refrained from making any judgements and drawing any conclusions from them. Before I attempt to do so, I would like to summarize some of their common characteristics, which I venture to believe, are quite general:

a. All processes that have been discussed are gauge-invariant.
b. The infrared divergences appear in color singlet channels.[12]
c. The perturbation theory results of these processes are equal to those corresponding to analogous QED processes.[13,14]
d. With the exception of the proof of the "weak KLN" theorem, while the QED results are well understood in general grounds and can be extended to all orders, the QCD results are obtained after doing hard calculations and no general arguments are, as of yet, known to extend them.
e. No apparent signal of confinement is present. Particularly the discussion of sections 4 and 5 indicates that there is a finite cross section for the process $\gamma \to q\bar{q}$ + gluons.

Does this mean that there is no confinement in QCD? Or does it just imply that perturbation theory is insufficient to tell us anything about confinement? Since I believe in QCD and in confinement, I suspect that the trouble lies more with perturbation theory and/or how one uses it. Succinctly and, necessarily simplistically, I would then like to state the situation as follows:

1. Either the perturbative approach has been improperly used, so that a proper reorganization of the expansion might still in fact yield a signal for confinement, or
2. Perturbation theory is incompatible with the notion of confinement. Non-perturbative methods must be sought to solve the problem.

In what follows, I would like to discuss what I have in mind regarding the first possibility. Of course I do not have, as yet, any concrete and/or workable alternative. I will just take the liberty to speculate, on the grounds of my own physical intuition and theoretical prejudice, about what could be plausible alternatives. I will conclude by discussing the second possibility.

A. Feeling the Long-Distance Charge and the "Skeleton" Expansion

In order to make the perturbation expansions sensible to start with, in the discussed processes, one defines a small coupling constant by a convenient choice of the renormalization point. This procedure controls large U.V. logarithms, but simultaneously it also seems to soften the IR ones. Order by order in perturbation theory they could give at most logarithmic modifications to the fundamental vertices. Even though the perturbation expansion could itself start to diverge, the overall order to order infrared structure will remain, perforce, the same. For example, this is the message from the proof of the weak KLN theorem, to wit Eq. 7: As long as ε is infinitesimal, the theorem applies. If ε is allowed to be finite (the infrared logs pile-up to something divergent like a power or worse) the theorem can fail. As I mentioned at the beginning, it is the ensemble of the IR logarithms, as regulated by Eq. (1), that presumably will tell us something about the confining mechanism. Clearly, the assumption that the fundamental vertices of the theory behave according to naive scaling laws at long distances, and that therefore the effective coupling is constant (up to logarithmic modifications) is probably inconsistent.

Recently, though, I have noticed that one could determine $\beta(g(k^2))$, $k^2 \to 0$, in the quark-antiquark-gluon sector of the theory, by just analyzing its infrared regime.[15] As the quarks and gluons approach their mass-shell, the infrared logs thus generated by radiative corrections add up to form the invariant charge (see Fig. 9).

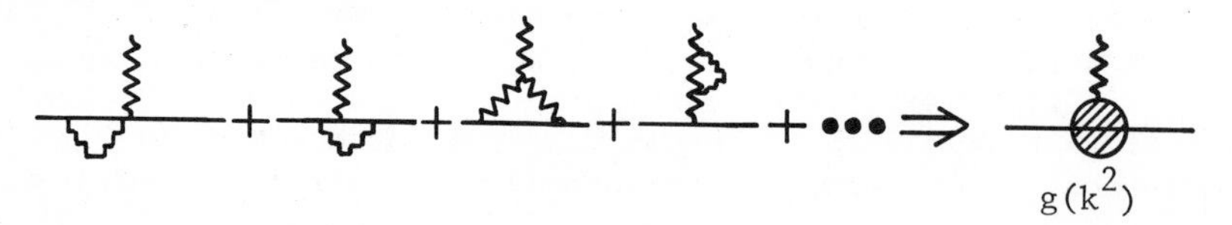

Figure 9

With this in mind, I would like to propose a reorganization of the graphical expansion so that the long distance aspects of the theory could become more apparent.[15,16] Namely, take any skeleton graph of the theory. Then add up all self-energy and vertex corrections so as to form the invariant vertices of the theory. The IR regime of these "dressed skeleton" graphs will become parametrized by the effective long distance charge and this behavior will then regulate the overall IR structure. Unfortunately, I have not been able yet to come up with a theoretically sound construction of this expansion. It is nevertheless interesting to see how dramatic changes in the IR structure of the theory could arise from it.

B. Threshold Behavior, Revisited

For example, the threshold behavior of $\pi(s)$ could become radically different. In perturbation theory, the naive potential that binds heavy quarks into "charmonia" is just g_μ^2/k^2 where g_μ^2 is the chosen running coupling, due to the one gluon-exchange kernel. (See Fig. 10). All higher order kernels give,

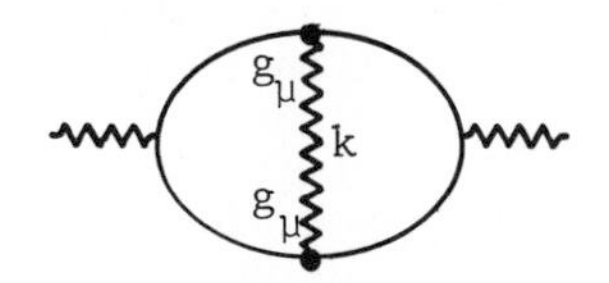

Figure 10

at worst, order by order, logarithmic modifications. The potential is effectively Coulomb. If one now re-organizes the kernel expansion into dressed skeletons (see Fig. 11), its contribution

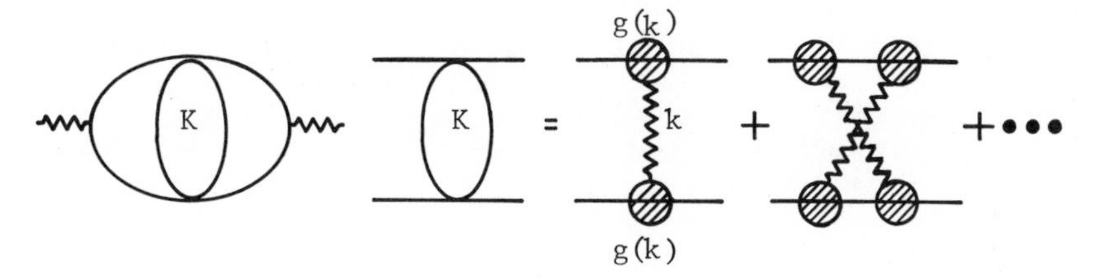

Figure 11

would be of the form $(g^2(k^2)/k^2)K[g(k^2)]$, where $K[g(k^2)]$ is some integral density in $g(k^2)$. This potential could clearly be quite different than Coulombic, in particular for $g^2(k^2)K[g(k^2)]\sim 1/k^2$ it would represent a linear binding potential.

C. Bloch-Nordsieck, Revisited: The Cornwall-Tiktoupolos Mechanism

Another situation where I could envision some new effect to appear is in the virtual-real infrared divergence cancellation program. In a dressed-skeleton-expansion formalism we would now have to compare, for example, the I.R. singularities coming from the diagrams in Fig. 12. The blobs represent the clustering of all radiative corrections. In the regions of integration that generate IR divergences the blobs approach $g(k^2)$, $k^2 \to 0$. It is clear that if $g(k^2)$ is no worse than logarithmically behaved, the standard cancellation will occur as usual (see for example

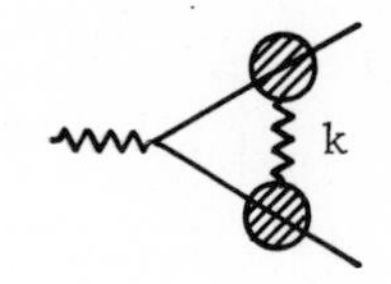

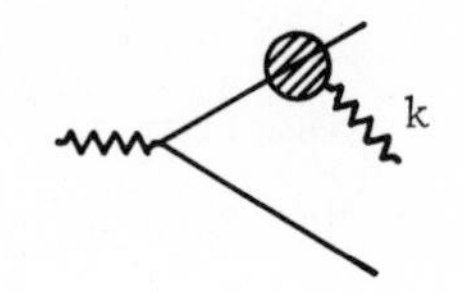

Figure 12

J. Carazzone's talk at these proceedings). If $g(k^2)$ is more singular, the gluon exchange diagram will have as usual a divergent contribution due to the $1/k^2$ of the gluon propagator that cancels with the divergent contribution from the emission graph. Furthermore, the exchange graph, because of the structure of $g(k^2)$ can give additional divergent contributions that cannot be cancelled.[17] This is certainly a plausible scenario for a breakdown of the KLN theorem. That this mismatch between real and virtual emission divergences can give a signal for confinement has already been suggested by Cornwall and Tiktopoulos (C.T.).[3] Specifically they considered the emission graph of Fig. 12 where the blob is the sum of leading-logarithms discussed in the first section. Assuming that the leading logs exponentiate into a damped exponential, the contribution to the cross-section from the sum over all possible emissions can be shown to be constant instead of a divergent power as in QED. Assuming then that the virtual diagrams behave as in QED, that is, they vanish as a power, the net amplitude for quark production will vanish. Even though I have some theoretical prejudice against various elements of the C.T. analysis (leading-log. invariant charge vs. renormalization group charge; the handling of virtual emission), I believe that their mechanism, confinement signal from the imbalance of virtual-real divergences, deserves further study.

D. Giving Up Perturbation Theory?

If and when one finds a workable "dressed-skeleton" expansion one would still be faced with some serious problems: what[16] is $g(k^2)$, i.e., $\beta(g(k^2))$, as $k^2 \to 0$? How will confinement and the confining mechanism be manifested?

Quite probably one will never be able to answer these questions within a purely perturbative framework. Of course perturbation theory, as we know it, may be truly incompatible with confinement ideas. I suspect, though, that one must still understand to deeper levels certain physical aspects of QCD, such as its vacuum structure[19], before one can really make real progress, either by perturbative or non-perturbative methods.

REFERENCES

1. T. W. Appelquist and J. Primack, Phys. Rev. D4, 2444 (1971) and references therein.

2. J. Carazzone, E. C. Poggio and H. R. Quinn, Phys. Rev. D11, 1228 (1975), and Phys. Rev. D12, 3368 (1975).

3. J. M. Cornwall and G. Tiktopoulos, Phys. Rev. D13, 3370 (1976) and Phys. Rev. Letts. 35, 338 (1975). Similar results have been obtained by R. Cahalan and J. Knight (unpublished) and Wu Yong-Shi and Dai Yuan-ben, Scientia Sinica 19, 65 (1975).

4. A comprehensive review of previous works and a compact rederivation of this result is given in the first part of reference 5.

5. E. C. Poggio and H. R. Quinn, Phys. Rev. D12, 3279 (1975).

6. T. W. Appelquist and H. D. Politzer, Phys. Rev. D12, 1404 (1975).

7. E. C. Poggio, H. R. Quinn and S. Weinberg, Phys. Rev. D13, 1958 (1976). See also R.Shankar, Harvard preprint, June 1976.

8. We are here loosely stating the overall message implied by the following works: a) T. Kinoshita, J. Math. Phys. 3, 650 (1962); b) T. D. Lee and M. Nauemberg, Phys. Rev. 133, B1544 (1964); c) T. Kinoshita and A. Ukawa, Phys. Rev. D13, 1573 (1976). J. Carazzone, in these proceedings, has discussed in more detail the application of b) to QCD.

9. E. C. Poggio and H. R. Quinn, Phys. Rev. D15, July (1976).

10. See, for example, S. Weinberg, Phys. Rev. 140B, 516 (1965).

11. T. W. Appelquist, J. Carazzone, H. Kluberg-Stern, and M. Roth, Phys. Rev. Letts. 36, 768 (1976). See also Y. P. Yao, Phys. Rev. Letts. 36, 542 (1976) and G. Sterman, University of Illinois preprints, to be published.

12. Notice, though, that the exponentiation result in reference 2 also holds when the incoming vector particle is a gluon.

13. T. W. Appelquist has called to my attention a paper by C. T. Sachradja, SLAC PUB. 1743 (Ap. 76), where he analyses the process $qq \to qq + 1$ gluon. Unlike the analogous QED process, $e^-e^- \to e^-e^- + 1$ photon,where the cross-section is linearly divergent. This result does not seem to have characteristic b).

14. E.de Rafael and K. Althes, Marseille preprint compute the quark-magnetic moment to be infrared divergent in second order.

15. E. C. Poggio, Phys. Rev. Letts. 36, 1511 (1976).

16. See also T. W. Appelquist, proceedings of the 1976 Coral Gables Conference and Yale University preprint.

17. I thank J. Carazzone for valuable discussions on this point.

18. See though P. Olesen, Niels Bohr preprint and proceedings of this conference.

19. F. Wilczek, these proceedings.

What Can We Learn About Quark Binding from Perturbation Theory?

JAMES CARAZZONE
Fermi National Accelerator Laboratory
Batavia, Illinois

Perturbation theory certainly cannot be expected to provide us with any reliable detailed information about an inherently strong coupling problem such as quark binding. Nevertheless, we might reasonably ask whether or not perturbation theory provides us with any hints about an underlying quark binding mechanism when applied to the "standard" model of quarks interacting with non-Abelian vector gluons (Quantum Chromodynamics or Q.C.D.).

In an attempt to investigate this question, Tom Appelquist, Hanna Kluberg-Stern, Mike Roth and myself set out to examine the classic Bloch-Nordsieck program[1,2,3] in the framework of Q.C.D. Let me remind you that in Quantum Electrodynamics the Bloch-Nordsieck program assures us that the infrared divergences associated with virtual corrections are cancelled by corresponding divergences in the emission of undetected photons whose total energy is less than the energy resolution ΔE of the detector. For the vertex in second order Q.E.D. this means that when we form a partial cross section the infrared divergences of the virtual exchange diagram:

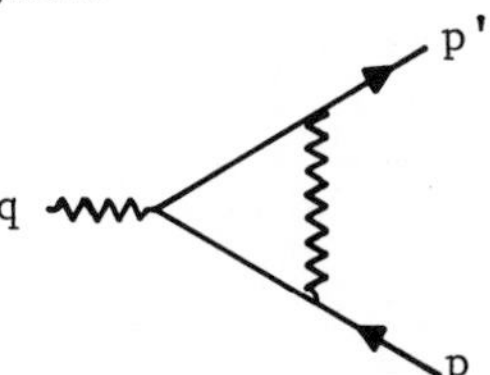

are cancelled by the divergence of the emission diagram

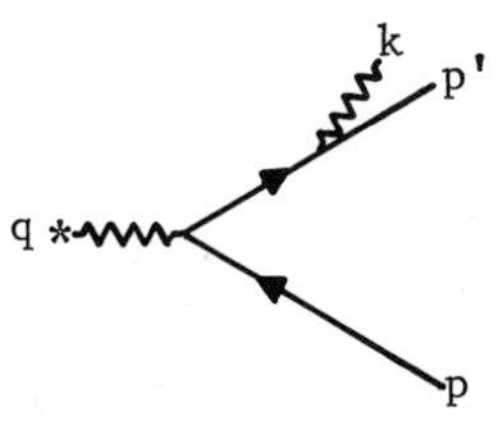

when the undetected emitted photon has energy less than the

energy resolution of the detector. For $q^2 > (2m_e)^2$, where m_e is the electron mass, the fact that the partial cross section is free of infrared divergence implies that in Q.E.D. it is possible to produce an electron-positron pair and to detect either the electron or positron at some distant point provided we recognize the fact that the process involves undetectable radiation of soft photons.

In Q.C.D. we want to examine essentially the same process--production of a quark-antiquark pair by a color singlet current,

$$J_\mu(x) = \sum_i \bar{q}_i(x)\, \gamma_\mu\, q_i(x) \ .$$

Instead of accompanying real and virtual photon radiation as we had in the Q.E.D. case, we now have gluon radiation in the Q.C.D. case. The detector has a finite energy resolution ΔE, and in order to obtain a gauge invariant cross section we assume the quark detector to trigger on some aspect of quark flavor. The quark detector therefore produces a sum over color states. Our experiment then looks like

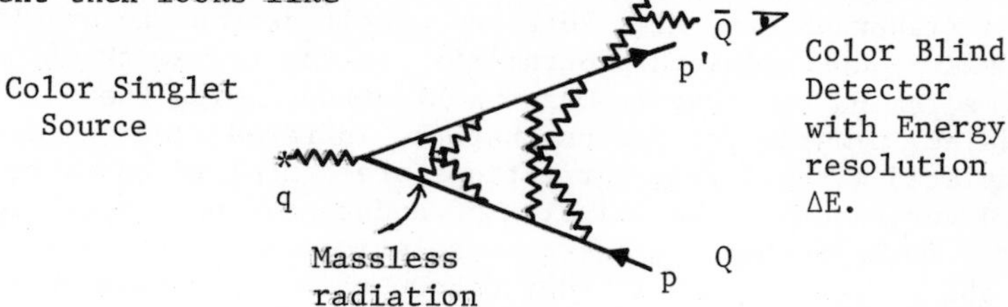

where $q^2 > (2\mu)^2$ for μ=non-zero quark mass.

Some differences with Q.E.D. exist of course. The most important of these is that in Q.C.D. the massless fields couple to themselves. This means that the coupling must be normalized at some off-mass-shell momentum point $p^2=-M^2$ so that the renormalized coupling is denoted by $g(M)$. Another difference with Q.E.D. is the nature of the infrared cutoff. The device of using a photon mass as a cutoff cannot be extended to field theories with a non-Abelian gauge invariance. The best way to cutoff the theory and maintain gauge invariance is to dimensionally continue to $4+\varepsilon$ dimensions where ε is complex and is taken to zero at the end.

The object to be computed is a unitless transition probability appropriate to the experiment described earlier,

$$R_{\Delta E}\left(\frac{E}{M}, \frac{\Delta E}{M}, \frac{\mu}{M}, \varepsilon, g(M)\right) \propto \int_{\Delta E} dE \left(\frac{d\sigma}{dE}\right) ,$$

where $R_{\Delta E}$ is normalized by the Born amplitude and E represents the energies (center of mass, quark energy). The calculation is organized by grouping together the different unitarity cuts of each diagram contributing to

$$\pi_{\mu\nu}(q) = \int (dx)\, e^{iq\cdot x} \langle 0|T(J_\mu(x)J_\nu(o))|0\rangle \ .$$

The phase-space integral over the quark momentum is restricted by the detector kinematics which include a finite energy resolution. For simplicity all calculations are carried out in Feynman gauge. The final result is independent of gauge choice.

On the two-loop level $\pi_{\mu\nu}(q)$ contains only ordinary Q.E.D. diagrams so that $R_{\Delta E}$ is clearly finite. It is instructive to repeat the argument for finiteness in the simple case of the vertex correction contribution to $\pi_{\mu\nu}(q^2)$:

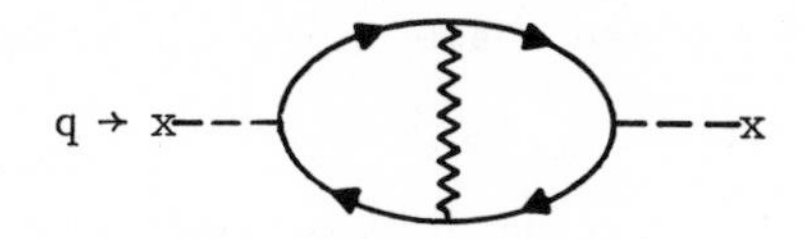

Altogether there are four unitarity cuts of this diagram. The following two:

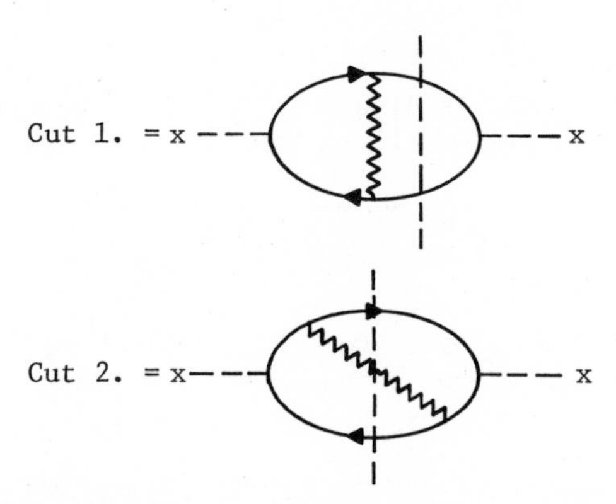

plus two others which are reflections of 1 and 2. The rules for unitarity cuts are:

1. A propagator factor of $i/p^2-m^2+i\varepsilon$ is replaced by $2\pi\delta_+(p^2-m^2)[2\pi\delta_-(p^2-m^2)]$ when the momentum runs from left to right (right to left) across the cut.

2. By convention, every factor to the right of a cut is complex conjugated.

In order to see the cancellation of infrared divergences between cuts 1 and 2 it is necessary to perform only the (dk_0) integration for cut 1. This is done by complex integration in the complex k_0 plane. Closing in the lower complex k_0 plane produces two possible infrared divergent contributions, from the pole due to the quark propagator $1/k^2+2p\cdot k+i\varepsilon$ we get,

$$[(2\pi)^2 C_N i g^2](-2\pi i)\int\!\!\int\frac{(d\vec{K})}{(2\pi)^4}F(p,p')\mathrm{Tr}[\cdots]\times$$

$$\times\frac{1}{2\sqrt{①}}\;\frac{1}{(-p_o+\sqrt{①})^2-\vec{K}^2-2p_o(-p_o+\sqrt{①})-2\vec{p}\cdot\vec{K}}\;\frac{1}{(-p_o+\sqrt{①})^2-\vec{K}^2},$$

where I have combined phase-space integrations appropriate to the detector as

$$\int F(p,p')\equiv\int\frac{(dp)}{(2\pi)^4}\int\frac{(dp')}{(2\pi)^4}\left\{\delta(q-p+p')\delta_+(p^2-\mu^2)\delta_-(p'^2-\mu^2)\right|_{\text{Detector}}$$

and in the center of mass system,

$$p_o = -p_o' = \sqrt{\vec{p}^2+\mu^2}$$

$$\sqrt{①} = \sqrt{(\vec{p}+\vec{K})^2+\mu^2}\ .$$

The integration over $(d\vec{K})$ is divergent for small $\vec{K}$, however the divergent piece of the integral can be written as

$$\int(d\vec{K})\;\frac{-1}{2p_o\left(\frac{\vec{p}\cdot\vec{K}}{p_o}\right)+2\vec{p}\cdot\vec{K}}\;\frac{1}{\left(\frac{\vec{p}\cdot\vec{K}}{p_o}\right)^2-\vec{K}^2},$$

which is odd in $\vec{K}\to-\vec{K}$ and therefore zero. The other infrared divergent contribution to cut 1 comes from the pole due to the $1/k^2+i\varepsilon$ factor,

$$[(2\pi)^2 C_N i g^2](-2\pi i)\int\!\!\int\frac{(d\vec{K})}{(2\pi)^4}F(p,p')\mathrm{Tr}[\cdots]\times$$

$$\times\frac{1}{2|\vec{K}|}\;\frac{1}{2p_o|\vec{K}|-2\vec{p}\cdot\vec{K}}\;\frac{1}{-2p_o|\vec{K}|-2\vec{p}\cdot\vec{K}}\ .$$

The integral of cut 2 can be evaluated directly as

$$-(2\pi)^3 C_N g^2 \iint \frac{(d\vec{K})}{(2\pi)^4} F(p,p'^2+2p'\cdot K)\mathrm{Tr}[\cdots] \times$$

$$\times \frac{1}{2|\vec{K}|} \frac{1}{2p_o|\vec{K}|-2\vec{p}\cdot\vec{K}} \frac{1}{2(-|\vec{K}|-\sqrt{①})|\vec{K}|-2\vec{p}\cdot\vec{K}},$$

the divergent piece of this integral exactly cancels the remaining divergent integral of cut 1.

In addition to the vertex correction contribution to $\pi_{\mu\nu}(q^2)$ we also have a self-energy insertion correction on the two loop level,

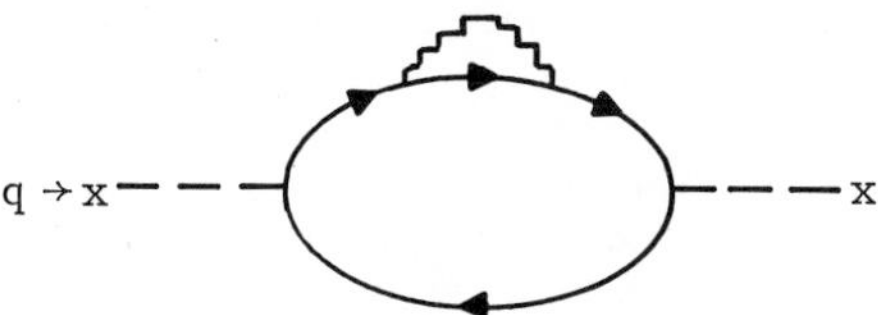

This diagram by itself is not infrared divergent, however when it is combined with the quark mass counter-term it becomes infrared divergent if it is subtracted at the quark mass-shell. Fortunately, our previous argument can be applied to the infrared divergent combination if we replace $1/\not{p}'-\mu+i\varepsilon$ by $1/\not{p}-\mu+i\varepsilon$ throughout the calculation. The result is the same: all infrared divergences cancel among the cuts of the diagram. Thus at the two loop level all infrared divergences cancel in an appropriately defined cross-section. Notice that if our detectors could distinguish color, we could not combine the various cuts as we did. We would be left with an infrared divergence.

On the three-loop level there are enormous number of diagrams to consider. It turns out that all but four are found in ordinary Q.E.D. and thus are known to have infrared divergences which cancel between the various cuts of each diagram.[4] The four new diagrams which are unique to Q.C.D. are

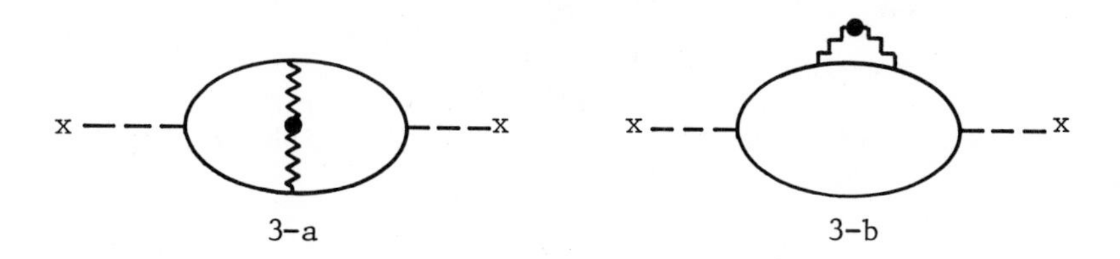

3-a 3-b

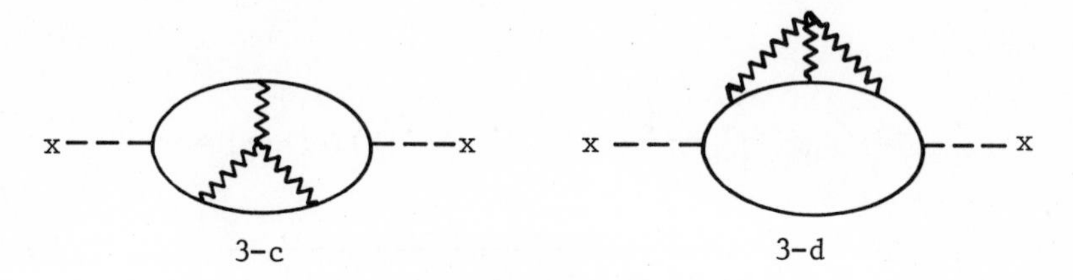

3-c 3-d

where

denotes the one-loop corrections to the gluon propagator. This object produces new infrared divergences not found in Q.E.D. because of the self-coupling of the massless gluon fields.

It turns out that by using the dispersion relation for the gluon propagator we can reduce the first two diagrams to the two-loop situation. Consider 3-a: the complete gluon propagator may be written as

$$(-i)\left(g_{\mu\nu}-\frac{k_\mu k_\nu}{k^2}\right)\frac{1}{k^2+i\varepsilon}\,d(k^2)\ ,$$

where $d(k^2)$ satisfies (see below)

$$\frac{d[\frac{k^2}{M^2},\varepsilon,g(M)]}{k^2+i\varepsilon}=\frac{d[0,\varepsilon,g(M)]}{k^2+i\varepsilon}+\int_0^\infty d\lambda^2\,\frac{\pi(\lambda^2)}{k^2-\lambda^2+i\varepsilon}\ .$$

The sum over unitarity cuts (with proper restrictions for the detector) of the first diagram is

$$\sum_{\Delta E\ \text{cuts}}\pi_{\mu\nu}^{(3-a)}(q)=d^{(2-\text{loop})}[0,\varepsilon,g(M)]\sum_{\Delta E\ \text{cuts}}\pi_{\mu\nu}^{(2-\text{loop})}(q,0)$$
$$+\int_0^\infty d\lambda^2\,\pi(\lambda^2)\sum_{\Delta E\ \text{cuts}}\pi_{\mu\nu}^{(2-\text{loop})}(q,\lambda)\ ,$$

where $\pi_{\mu\nu}^{(2-\text{loop})}(q,\lambda)$ corresponds to the two-loop vertex correction diagram considered earlier except with a mass λ for the internal vector boson. The argument given earlier shows that the sum over cuts is finite even for $\lambda=0$. Thus the above equation is free of infrared divergence as $\varepsilon\to 0$ provided

$$d^{(2-\text{loop})}[0,\varepsilon,g(M)]+\int_{\to 0}d\lambda^2\,\pi(\lambda^2),$$

is finite. This is indeed the case because $d[k^2/M^2,\varepsilon,g(M)]$ is finite as $\varepsilon \to 0$ for $k^2/M^2 \neq 0$ since it is normalized at $k^2=-M^2$. This argument has only one slight complication: the dispersion integral cannot really be taken down to $\lambda=0$ because the origin is an essential singularity. The same conclusion follows, however, if use is made of a small circle of radius δ about the origin in the complex λ^2 plane.

The next two diagrams, 3-c and 3-d are much more difficult to deal with. Just as in the two-loop case the sum of cuts of diagram 3-d can be proven to be free of divergences by essentially the same argument that applies to 3-c. The same kind of analysis that I used in the two-loop case must be applied to 3-c with the added complication that there are now angular infrared divergences that are unique to theories of self-coupled massless gauge fields.

To begin, we want to consider the diagram of 3-c with rootings as follows:

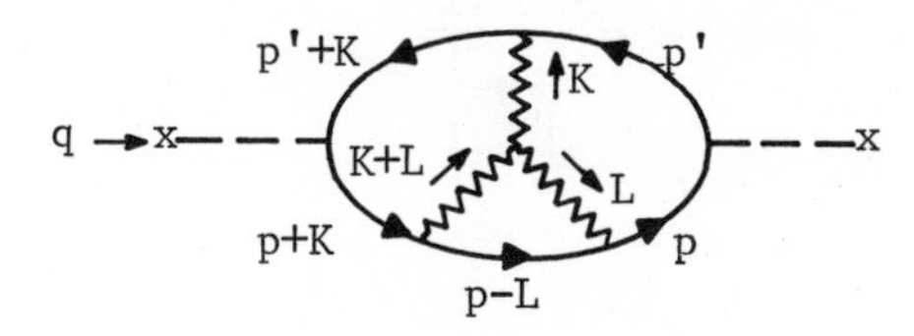

Altogether there are ten cuts of this diagram. The following five plus five more which are just reflections of the first five:

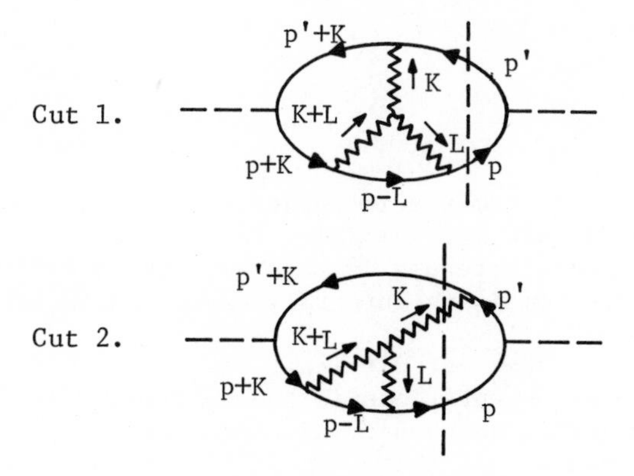

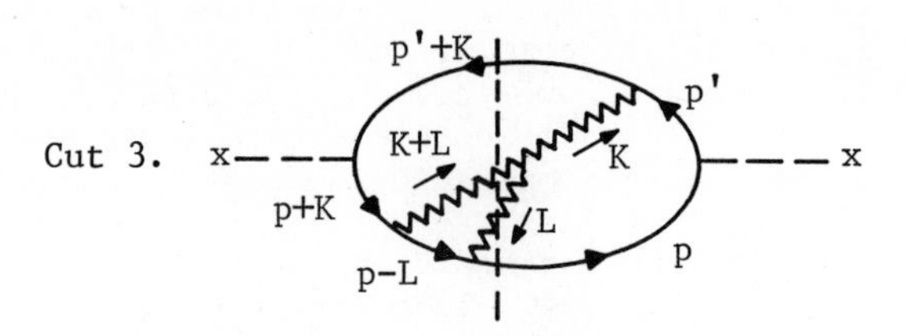

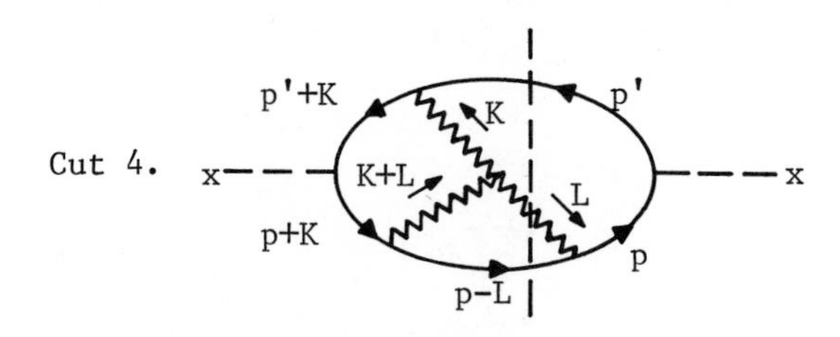

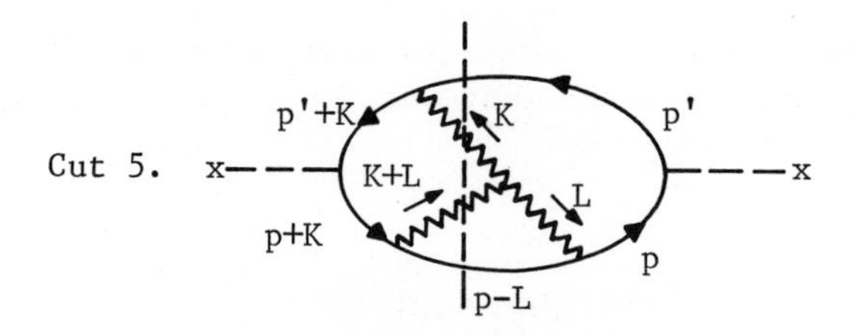

The procedure is to perform the integrations over (dk_o) and $(d\ell_o)$ and then to isolate all remaining infrared divergences. The proof of infrared finiteness is to show that as $\varepsilon \to 0$ the divergent parts of the remaining integrals cancel among these five different cuts.

The first cut is the most difficult because both integrations must be done as integrations in the complex plane. First let me make the notation as compact as possible:

Each diagram contains a trace factor of

$$\mathrm{Tr}[\cdots]^{\mu\nu\lambda} \equiv \mathrm{Tr}[\gamma_\sigma(\not{p}+\not{K}+\mu)\gamma^\mu(\not{p}-\not{L}+\mu)\gamma^\nu(\not{p}+\mu) \times$$
$$\times\ \gamma^\sigma(\not{p}'+\mu)\gamma^\lambda(\not{p}'+\not{K}+\mu)]$$

and a tri-vector vertex factor of

$$\Gamma^{\mu\nu\lambda} \equiv (2L+K)^\lambda g^{\mu\nu} + (K-L)^\mu g^{\nu\lambda} - (2K+L)^\nu g^{\mu\lambda}$$

in the combination

$$N = \mathrm{Tr}[\cdots]^{\mu\nu\lambda}\Gamma_{\mu\nu\lambda}$$

inside of a phase-space integral

$$g^4(\text{Group theory factor}) \int \frac{(dp)}{(2\pi)^4} \int \frac{(dp')}{(2\pi)^4} \int \frac{dK^{n-1}}{(2\pi)^4} \int \frac{dL^{n-1}}{(2\pi)^4} N \times$$
$$\times\ (\delta(q-p-p')(2\pi)^2\delta_+(A^2-\mu^2)\delta_-(B^2-\mu^2)\Big|_{\text{detector}}$$
$$\equiv \int F(A,B),$$

where A and B are quark momentum vectors. By $n=4+\varepsilon$ I mean that the integrals are regulated by dimensional continuation with (complex) $\varepsilon\to 0$ at the end. By $(\cdots|_{\text{detector}}$ I mean that the detector restrictions must also be inserted, i.e., the total energy of any emitted massless quanta is less than ΔE. In addition there may be angular restrictions applied to the detected quark. Then the diagram of cut 1 is

$$C_{1.} = \int F(p,p') \int_{K_o,\ell_o} \frac{1}{K^2+2p\cdot K+i\varepsilon}\ \frac{1}{K^2+2p'\cdot K+i\varepsilon}\ \frac{1}{\ell^2-2p\cdot\ell+i\varepsilon} \times$$
$$\times \frac{1}{K^2+i\varepsilon}\ \frac{1}{\ell^2+i\varepsilon}\ \frac{1}{(K+\ell)^2+i\varepsilon}\ ,$$

where $\int_{K_o} = \int_{-\infty}^{+\infty} dK_o/2\pi$ etc. If I want to think of doing the $d\ell_o$ integration first I may write this as a sum of three terms

$$C_{1.} = C_{1.}^{1.} + C_{1.}^{2.} + C_{1.}^{3.} = \int F(p,p') \int_{K_o,\ell_o} (-2\pi i)\{\cdots\} \times$$
$$\times \frac{1}{K^2+i\varepsilon}\ \frac{1}{K^2+2p\cdot K+i\varepsilon}\ \frac{1}{K^2+2p'\cdot K+i\varepsilon}\ ,$$

where (in the complex ℓ_o plane)

$$\{\cdots\} = \delta_+(\ell^2-2p\cdot\ell+i\varepsilon)\ \frac{1}{2p\cdot\ell}\ \frac{1}{K^2+2K\cdot\ell+2p\cdot\ell}$$

$$+\ \delta_+(\ell^2+i\varepsilon)\ \frac{-1}{2p\cdot\ell}\ \frac{1}{K^2+2K\cdot\ell}$$

$$+\ \delta_+((K+\ell)^2+i\varepsilon)\ \frac{-1}{K^2+2K\cdot\ell+2p\cdot\ell}\ \frac{-1}{K^2+2K\cdot\ell}\ .$$

The delta functions force ℓ_o to have the values (in c.m. frame)

$$\ell_o = p_o + \sqrt{①} - i\varepsilon/2\sqrt{①} \qquad \text{in } C_{1.}^{1.}$$

$$\ell_o = |\vec{\ell}| - i\varepsilon/2|\vec{\ell}| \qquad \text{in } C_{1.}^{2.}$$

$$\ell_o = -K_o + |\vec{K}+\vec{\ell}| - i\varepsilon/2|\vec{K}+\vec{\ell}| \qquad \text{in } C_{1.}^{3.}$$

$$\sqrt{①} = \sqrt{(\vec{p}-\vec{\ell})^2 + \mu^2}\ .$$

I must be careful to retain the $i\varepsilon$ terms as $\varepsilon\to 0^+$ because this generates restrictions on which poles I pick up when I do the dk_o integration.

Next I do the dk_o integration and let $\varepsilon\to 0^+$ in the denominators. For $C_{1.}^{1.}$ there are four poles in the complex k_o plane which I label by means of a second index:

$$C_{1.}^{1.} = C_{1.}^{1.,1.} + C_{1.}^{1.,2.} + C_{1.}^{1.,3.} + C_{1.}^{1.,4.}$$

The first is at $k_o=|\vec{k}+\vec{\ell}|-\sqrt{①}-p_o$ and has no infrared singularities of any kind. The second is at $k_o = |\vec{k}|$ and can be written as

$$(-2\pi i)^2 \int F(p,p')[\cdots]$$

where [---] is obtained by substitution of $\ell_o=p_o+\sqrt{①}$ and $k_o=|\vec{k}|$ in the above expressions. Inspection shows that this term has a D=0 infrared divergence in the $d\vec{k}$ integration as $|\vec{k}|\to 0$ for $|\vec{\ell}|\neq 0$ and no other divergence. I shall label this source of divergence $C_{1.}^{1.,2.}(k_s\ell_h,0)$ where the subscript stands for cut 1., and the two superscripts denote the particular term. The two variables describe the limit which generates the infrared divergence, and the zero means that the divergence is logarithmic. Clearly a single term may generate several infrared divergences each from different regions of the integration variables. To denote this I will express the contribution of a given term as sum of terms each one of which has an infrared divergence arising from a particular limit of the integration variables. Finally I denote the degree of divergence by the last variable. The term described above as $C_{1.}^{1.,2.}(k_s\ell_h,0)$ will eventually be canceled by a term in cut 2 referred to as $C_{2.}^{1.}(k_s\ell_h,0)$ for which $k_o=|\vec{k}|$ and $\ell_o=p_o+\sqrt{①}$.

The third pole in the k_o plane is at

$$k_o = -p_o + \sqrt{②} ,$$

where

$$\sqrt{②} = \sqrt{(\vec{p}+\vec{K})^2 + \mu^2} .$$

Remember all $C_1^{1.}$ terms have $\ell_o=p_o+\sqrt{①}$. This term has a D=0 divergence for $k_s\ell_h$ but does not contribute in this limit because it is odd under $\vec{k}\to-\vec{k}$. Thus $C_{1.}^{1.,3.}(k_s\ell_h,0)=0$. The term $C_{1.}^{1.,4.}$ is at $k_o=p_o+\sqrt{②}$ which is bounded away from the origin in the k_o plane; thus there is no infrared divergence.

For the term $C_{1.}^{2.}(\ell_o=|\vec{\ell}|)$ we begin to encounter a more complex situation. There are three poles in the complex k_o plane which produce infrared divergence. The first one of these, $k_o=|\vec{k}+\vec{\ell}|-|\vec{\ell}|$, produces a very complicated term

$$C_{1.}^{2.,1.} = C_{1.}^{2.,1.}(k_s\ell_h,-1) + C_{1.}^{2.,1.}(k_s\ell_s,0) + C_{1.}^{2.,1.}(\vec{k}||\vec{\ell},0),$$

where $C_{1.}^{2.,1.}(\vec{k}||\vec{\ell},0)$ denotes an angular logarithmic divergence which occurs for $\vec{k}$ parallel to $\vec{\ell}$. Fortunately the entire term, $C_{1.}^{2.,1.}$, is identically canceled by the term $C_{1.}^{3.,3.}$ which has the same values of $\ell_o=|\vec{\ell}|$ and $k_o=|\vec{k}+\vec{\ell}|-|\vec{\ell}|$. At $k_o=|\vec{k}|$ we encounter a term, $C_{1.}^{2.,2.}$, which contains the factor

$$\frac{1}{2|\vec{K}||\vec{\ell}| - 2\vec{K}\cdot\vec{\ell}} ,$$

which generates an angular logarithmic divergence when $\vec{k}$ is parallel to $\vec{\ell}$ (essentially the same term generates $C_{1.}^{2.,1.}(\vec{k}||\vec{\ell},0)$). Altogether we have the following infrared divergence structure:

$$C_{1.}^{2.,2.} = C_{1.}^{2.,2.}(k_s\ell_h,-1) + C_{1.}^{2.,2.}(k_h\ell_s,0)$$
$$+ C_{1.}^{2.,2.}(k_s\ell_s,0) + C_{1.}^{2.,2.}(\vec{k}||\vec{\ell}.0).$$

The piece $C^{2.,2.}_{1.}(k_s\ell_h,-1)$ combines with $C^{3.,4.}_{1.}(k_s\ell_h,-1)$ (both have $k_o=|\vec{k}|$ and $\ell_o = |\vec{\ell}|$ in this limit) to produce a remainder, $[C^{2.,2.}_{1.}(k_s\ell_h,-1) + C^{3.,4.}_{1.}(k_s\ell_h,-1)]$ which has only a D=0 divergence. The remainder finally cancels with a similar combination of terms in cut 2., $[C^{2.}_{2.}(k_s\ell_h,-1) + C^{3.}_{2.}(k_s\ell_h,-1)]$. The term $C^{2.,2.}_{1.}(k_h\ell_s,0)$ cancels with $C^{3.,2.}_{1.}(k_h\ell_s,0)$ when $\vec{\ell} \rightarrow -\vec{\ell}$ in the $k_h\ell_s$ limit. The "double soft" term $C^{2.,2.}_{1.}(k_s\ell_s,0)$ cancels against $C^{2.}_{2.}(k_s\ell_s,0)$, and the angular piece, $C^{2.,2.}_{1.}(\vec{k}||\vec{\ell},0)$ will be discussed later.

The last contributing pole in the k_o plane is at $k_o=-p_o+\sqrt{②}$ produces $C^{2.,3.}_{1.} = C^{2.,3.}_{1.}(k_s\ell_h,-1)+C^{2.,3.}_{1.}(k_s\ell_s,0)$. The term $C^{2.,3.}_{1.}(k_s\ell_h,-1)$ combines with $C^{3.,5.}_{1.}(k_s\ell_h,-1)$ to produce a D=0 remainder which is explicitly odd under $\vec{k} \rightarrow -\vec{k}$ and hence zero. The term $C^{2.,3.}_{1.}(k_s\ell_s,0)$ in which $|\vec{k}|$ and $|\vec{\ell}|$ both go to zero, can be seen to cancel $C^{3.,5.}_{1.}(k_s\ell_s,0)$ if we note that the transformation,

$$\vec{K} \rightarrow -\vec{K}'$$
$$\vec{\ell} \rightarrow \vec{\ell}' + \vec{K}',$$

leaves the numerator factor, N, invariant <u>in this limit only.</u>

Next we come to the terms of $C^{3.}_{1.}(\ell_o=-k_o+|\vec{k}+\vec{\ell}|)$ several of which I have already mentioned. In this case, there are five poles in the complex k_o plane which generate infrared divergences. Briefly we have:

1) $C^{3.,1.}_{1.}$ at $k_o=|\vec{k}+\vec{\ell}|-p_o+\sqrt{①}$; $\ell_o=p_o-\sqrt{①}$ with

$$C^{3.,1.}_{1.} = C^{3.,1.}_{1.}(k_h\ell_s,0) + C^{3.,1.}_{1.}(k_s\ell_s,0).$$

The term $C^{3.,1.}_{1.}(k_h\ell_s,0)$ is odd under $\vec{\ell} \rightarrow -\vec{\ell}$ and vanishes in the $k_h\ell_s$ limit. The term $C^{3.,1.}_{1}(k_s\ell_s,0)$ cancels with a term of cut 4., $C^{1.}_{4.}(k_s\ell_s,0)$, if we make the transformation

$$\vec{K} \rightarrow -\vec{K}'$$
$$\vec{\ell} \rightarrow \vec{\ell}' + \vec{K}' ,$$

in the term of cut 4.

2) $C^{3.,2.}_{1.}$ at $k_o=|\vec{k}+\vec{\ell}|+|\vec{\ell}|$; $\ell_o=-|\vec{\ell}|$ with

$$C^{3.,2.}_{1.} = C^{3.,2.}_{1.}(k_h\ell_s,0) + C^{3.,2.}_{1.}(k_s\ell_s,0) + C^{3.,2.}_{1.}(\vec{k}||\vec{\ell},0).$$

The term $C^{3.,2.}_{1.}(k_h\ell_s,0)$ cancels $C^{2.,2.}_{1.}(k_h\ell_s,0)$ as already mentioned. The term $C_{1.}^{3.,2.}(k_s\ell_s,0)$ cancels with the only term of cut 3. in the $k_s\ell_s$ limit, $C_{3.}(k_s\ell_s,0)$. The angular divergence I leave for later.

3) $C_{1.}^{3.,3.}$ at $k_o=|\vec{k}+\vec{\ell}|-|\vec{\ell}|$; $\ell_o=|\ell|$ cancels $C_{1.}^{2.,1.}$ as promised.

4) $C_{1.}^{3.,4.}$ at $k_o=|\vec{k}|$, $\ell_o=-|\vec{k}|+|\vec{k}+\vec{\ell}|$ with

$$C_{1.}^{3.,4.} = C_{1.}^{3.,4.}(k_s\ell_h,-1) + C_{1.}^{3.,4.}(k_s\ell_s,0) + C_{1.}^{3.,4.}(\vec{k}||\vec{\ell},0).$$

The term $C_{1.}^{3.,4.}(k_s\ell_h,-1)$ combines with $C_{1.}^{2.,2.}(k_s\ell_h,-1)$ as already discussed. The "double-soft" term $C_{1.}^{3.,4.}(k_s\ell_s,0)$ cancels a term of cut 2., $C_{2.}^{3.}(k_s\ell_s,0)$.

5) $C_{1.}^{3.,5.}$ at $k_o=-p_o+\sqrt{②}$; $\ell_o=-p_o+\sqrt{②}+|\vec{k}+\vec{\ell}|$ with

$$C_{1.}^{3.,5.} = C_{1.}^{3.,5.}(k_s\ell_h,-1) + C_{1.}^{3.,5.}(k_s\ell_s,0).$$

The $C_{1.}^{3.,5.}(k_s\ell_h,-1)$ term combines with $C_{1.}^{2.,3.}(k_s\ell_h,-1)$, and the $C_{1.}^{3.,5.}(k_s\ell_s,0)$ term cancels $C_{1.}^{2.,3.}(k_s\ell_s,0)$ (after transformation) as promised.

Our analysis of cut 1. is now complete, aside from angular divergences, and we turn to cut 2. which is much less complicated due to the presence of $\delta_+(k^2)$. Only three terms are present:

1) $C_{2.}^{1.}$ at $k_o=|\vec{k}|$, $\ell_o=p_o+\sqrt{①}$ with

$$C_{2.}^{1.} = C_{2.}^{1.}(k_s\ell_h,0).$$

This term cancels $C_{1.}^{1.,2.}(k_s\ell_h,0)$.

2) $C_{2.}^{2.}$ at $k_o=|\vec{k}|$, $\ell_o=|\vec{\ell}|$ with

$$C_{2.}^{2.} = C_{2.}^{2.}(k_s\ell_h,-1) + C_{2.}^{2.}(k_h\ell_s,0) + C_{2.}^{2.}(k_s\ell_s,0) + C_{2.}^{2.}(\vec{k}||\vec{\ell},0).$$

The term $C_{2.}^{2.}(k_s\ell_h,-1)$ combines with $C_{2.}^{3.}(k_s\ell_h,-1)$ to become only D=0 in this limit. The combination cancels $[C_{1.}^{2.,2.}(k_s\ell_h,-1) + C_{1.}^{3.,4.}(k_s\ell_h,-1)]$ as already explained. The term $C_{2.}^{2.}(k_h\ell_s,0)$ is canceled by the single term of cut 3. in this limit, $C_{3.}(k_h\ell_s,0)$. The double soft term $C_{2.}^{2.}(k_s\ell_s,0)$ cancels $C_{1.}^{2.,2.}(k_s\ell_s,0)$.

3) $C_{2.}^{3.}$ at $k_o=|\vec{k}|$, $\ell_o=|\vec{k}+\vec{\ell}|-|\vec{k}|$ with

$$C_{2.}^{3.} = C_{2.}^{3.}(k_s\ell_h,-1) + C_{2.}^{3.}(k_s\ell_s,0) + C_{2.}^{3.}(\vec{k}||\vec{\ell},0).$$

As just mentioned, $C_{2.}^{3.}(k_s\ell_h,-1)$ combines with $C_{2.}^{2.}(k_s\ell_h,-1)$, and $C_{2.}^{3.}(k_s\ell_s,0)$ cancels $C_{1.}^{3.,4.}(k_s\ell_s,0)$.

Cut 3. contains only one term because of the two delta functions. It is $C_{3.}$ at $k_o=|\vec{\ell}|+|\vec{k}+\vec{\ell}|$, $\ell_o=-|\vec{\ell}|$ with

$$C_{3.} = C_{3.}(k_h\ell_s,0) + C_{3.}(k_s\ell_s,0) + C_{3.}(\vec{k}||\vec{\ell};0) .$$

The term $C_{3.}(k_h\ell_s,0)$ cancels $C_{2.}^{2.}(k_h\ell_s,0)$ after we let $\vec{\ell}\rightarrow-\vec{\ell}$, and the term $C_{3.}(k_s\ell_s,0)$ cancels $C_{1.}^{3.,2.}(k_s\ell_s,0)$.

Cut 4. has a $\delta_+(\ell^2)$ so that $\ell_o=|\vec{\ell}|$ and there are three terms:

1) $C_{4.}^{1.}$ at $k_o=-p_o+\sqrt{②}$ with $C_{4.}^{1.}=C_{4.}^{1.}(k_s\ell_s,0)$ as the only divergence. This term cancels $C_{1.}^{3.,1.}(k_s\ell_s,0)$ after transformation.

2) $C_{4.}^{2.}$ at $k_o=|\vec{k}+\vec{\ell}|-|\vec{\ell}|$ with

$$C_{4.}^{2.} = C_{4.}^{2.}(k_h\ell_s,0) + C_{4.}^{2.}((k+\ell)_s(k-\ell)_h,0) + C_{4.}^{2.}(k_s\ell_s,0) + C_{4.}^{2.}(\vec{k}||\vec{\ell},0).$$

The term $C_{4.}^{2.}(k_h\ell_s,0)$ cancels with the next term, $C_{4.}^{3.}(k_h\ell_s,0)$; and $C_{4.}^{2.}((k+\ell)_s(k-\ell)_h,0)$ cancels with the same limit of the single term of cut 5., $C_{5.}((k+\ell)_s(k-\ell)_h,0)$. The term $C_{4.}^{2.}(k_s\ell_s,0)$ changes sign under

$$\vec{K}\rightarrow-\vec{K}'$$
$$\vec{\ell}\rightarrow\vec{K}'+\vec{\ell}'$$

in the $k_s\ell_s$ limit and is therefore zero.

3) $C_{4.}^{3.}$ at $k_o=|\vec{k}|$ with

$$C_{4.}^{3.} = C_{4.}^{3.}(k_s\ell_h,0) + C_{4.}^{3.}(k_h\ell_s,0) + C_{4.}^{3.}(k_s\ell_s,0) + C_{4.}^{3.}(\vec{k}||\vec{\ell},0).$$

The term $C_{4.}^{3.}(k_s\ell_h,0)$ cancels the single term of cut 5. in the limit, $C_{5.}(k_s\ell_h,0)$, if we let $\vec{k}\rightarrow-\vec{k}$. As just mentioned $C_{4.}^{3.}(k_h\ell_s,0)$ cancels with $C_{4.}^{2.}(k_h\ell_s,0)$, and $C_{4.}^{3.}(k_h\ell_s,0)$ is canceled by $C_{5.}(k_s\ell_s;0)$ if we make the transformation

$$\vec{K}\rightarrow-\vec{K}'$$
$$\vec{\ell}\rightarrow\vec{K}'+\vec{\ell}' .$$

Lastly, we have cut 5. which has only one term because of the two delta functions. These imply that

$$K_o = -|\vec{K}|$$
$$\ell_o = |\vec{K}| + |\vec{K}+\vec{\ell}| ,$$

and we find

$$C_{5.} = C_{5.}(k_s\ell_h,0) + C_{5.}((k+\ell)_s(k-\ell)_h,0)$$
$$+ C_{5.}(k_s\ell_s,0) + C_{5.}(\vec{k}||\vec{\ell},0).$$

As already explained I find that $C_{5.}(k_s\ell_h,0)$ cancels $C_{4.}^{3.}(k_s\ell_h,0)$; $C_{5.}((k+\ell)_s(k-\ell)_h,0)$ cancels $C_{4.}^{2.}((k+\ell)_s(k-\ell)_h,0)$, and $C_{5.}(k_s\ell_s,0)$ cancels $C_{4.}^{3.}(k_h\ell_s,0)$.

There now remains only the angular divergences which are easy to treat. Define,

$$\text{Cos }\theta = \frac{\vec{K}\cdot\vec{\ell}}{|\vec{K}||\vec{\ell}|} ,$$

for $|\vec{k}|\neq 0$ and $|\vec{\ell}|\neq 0$, and I consider cut 1. in detail first. There are three terms with angular divergence: $C_{1.}^{2.,2.}(\vec{k}||\vec{\ell},0)$, $C_{1.}^{3.,2.}(\vec{k}||\vec{\ell},0)$ and $C_{1.}^{3.,4.}(\vec{k}||\vec{\ell},0)$. Each of these terms contains logarithmic divergence in the angular integration over cos θ. It is not difficult to check that $C_{1.}^{2.,2.}(\vec{k}||\vec{\ell},0)$ has its divergence only at cos θ = 1 for any value of $|\vec{k}|$ and $|\vec{\ell}|$ not zero. The term $C_{1.}^{3.,2.}(\vec{k}||\vec{\ell},0)$ has divergence only when cos θ= -1 and $|\vec{k}|>|\vec{\ell}|>0$, and $C_{1.}^{3.,4.}(\vec{k}||\vec{\ell},0)$ has divergence at cos θ= 1 for $|\vec{k}|,|\vec{\ell}|>0$ and at cos θ =-1 for $|\vec{k}|>|\vec{\ell}|>0$. To denote these situations I write,

$$C_{1.}^{2.,2.}(\vec{k}||\vec{\ell},0) = C_{1.}^{2.,2.}(1)$$
$$C_{1.}^{3.,2.}(\vec{k}||\vec{\ell},0) = C_{1.}^{3.,2.}(-1,|\vec{k}| > |\vec{\ell}|$$
$$C_{1.}^{3.,4.}(\vec{k}||\vec{\ell},0) = C_{1.}^{3.,4.}(1) + C_{1.}^{3.,4.}(-1,|\vec{k}| > |\vec{\ell}|).$$

It is easy to show that $C_{1.}^{2.,2.}(1)$ cancels $C_{1}^{3.,4.}(1)$ and $C_{1.}^{3.,2.}(-1,|\vec{k}|>|\vec{\ell}|)$ cancels $C_{1.}^{3.,4.}(-1,|\vec{k}|>|\vec{\ell}|)$. Almost precisely the same analysis applies to the angular divergences of cuts 2. and 3. and to the divergences of cuts 4. and 5.

What can we learn from this long analysis? At least two things are clear:

1. It is essential that the infrared divergences be no stronger than they are in 4 dimensions. I expect the above

cancellation of infrared divergences to break down in two or three dimensions.

2. Because I had to use explicit properties of the numerator structure to show the cancellation, I assume that the cancellation may be interpreted as a direct result of the gauge invariance of the couplings.

Furthermore, if we are willing to assume that the cancellation continues to hold to all orders, we obtain the result that perturbation theory implies that it is indeed possible to separate quark flavor--hence no confinement.

REFERENCES

1. T. Appelquist, J. Carazzone, H. Kluberg-Stern, and M. Roth, Phys. Rev. Lett. 36, 768 (1976).
2. F. Bloch and A. Nordsieck, Phys. Rev. 52, 54 (1937).
3. T. Kinoshita, J. Math. Phys. (N.Y.) 3, 650 (1962).
4. D. R. Yennie, S. C. Frautschi, and H. Suura, Ann. Phys. (N.Y.) 13, 379 (1961); also G. Grammer and D. R. Yennie, Phys. Rev. D8, 4332 (1973).
5. T. D. Lee and M. Nauenberg, Phys. Rev. 133, B1549 (1964).

A Fermionic Representation of the Infrared Behavior of Pure Yang-Mills Fields

POUL OLESEN
The Niels Bohr Institute
University of Copenhagen
Copenhagen, Denmark

I. INTRODUCTION AND THE MAIN RESULTS

In this talk I am going to present some work on the infrared behavior of pure Yang-Mills fields obtained during the last year.[1,2] Making certain assumptions (valid to any finite order of perturbation theory), to be discussed in detail in Sec. 2, we obtain the following three results:

(A) Consider a pure gluon theory

$$L_{Y-M} = -\frac{1}{2}\,\mathrm{Tr}\{(\partial_\mu B_\nu - \partial_\nu B_\mu - g[B_\mu, B_\nu])^2\} + \text{gauge terms}, \tag{1.1}$$

with no quarks. In this theory one can in principle calculate any Greens functions, for example

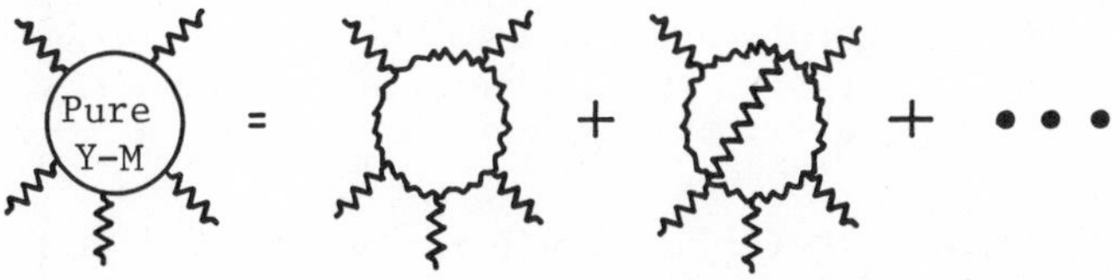

(1.2)

It then turns out that at large distances this is equivalent to a theory with dynamically active quarks,

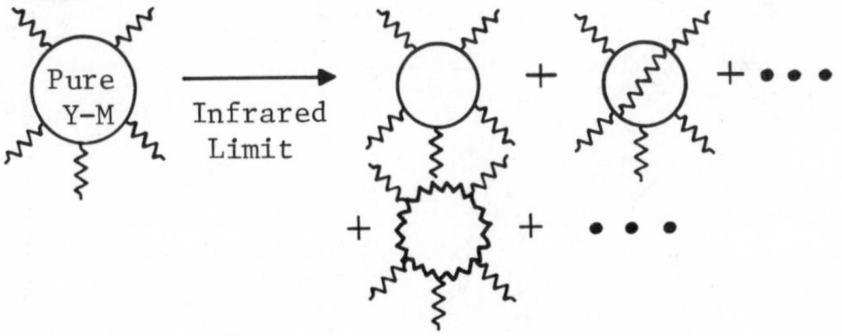

(1.3)

In (1.3) and (1.2) the right-hand sides are differently parametrized. In particular, in (1.3) we use Weinberg's zero-mass renormalization scheme.[3] Equations (1.2) and (1.3) show that in the infrared limit a theory without quarks becomes equivalent to a theory with quarks!

(B) In Eq. (1.3) the number of fermion multiplets (the flavor) is arbitrary. We then show that by taking "enough" fermions the right-hand side of Eq. (1.3) becomes dominated by diagrams which contain at least one fermion loop, i.e.,

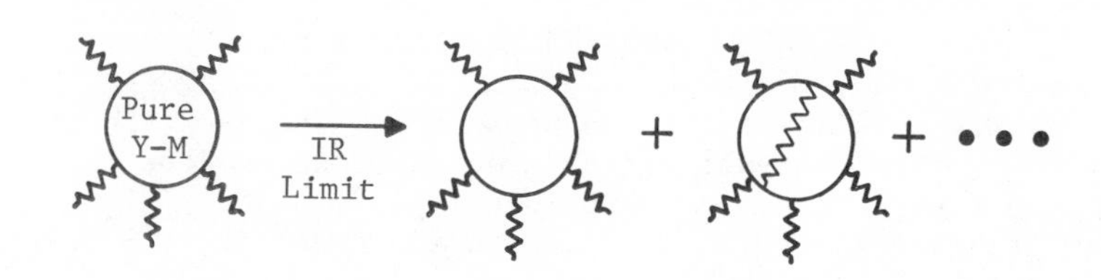

(1.4)

The pure Yang-Mills theory is presumably a strong coupling theory. However, the right-hand side of Eq. (1.4) is a weak coupling theory (remember that we need different parametrizations in (1.2) and (1.3)). The weak coupling is, however, paid by an infinite effective mass (in the sense of Weinberg[3]) of the quarks. In the following we refer to the right-hand side of (1.4) as the "fermionic representation" of the infrared behavior of the pure Yang-Mills theory. It appears to me to be an interesting question as to how the "spontaneously generated" quark pairs are to be interpreted physically, and whether this phenomenon implies that quarks are less "fundamental" than gluons.

(C) From the result in (B) it then turns out that one can show that the pure Yang-Mills Greens functions are power behaved

$$\Gamma^{(n)}(\lambda p) \sim \lambda^{4-n-nf_o}, \quad \lambda \to 0, \tag{1.5}$$

where $\Gamma^{(n)}$ is the n-point one-particle-irreducible Greens functions for the pure Yang-Mills theory. Furthermore, one can show that

$$f_o > 0, \tag{1.6}$$

so that the behavior (1.5) is more singular than the "canonical" behavior λ^{4-n}. It is interesting to compare Eq. (1.5) with the corresponding behavior in perturbation theory obtained by

Poggio and Quinn[4] (see also Poggio's talk at this workshop), namely (to any finite order)

$$\Gamma^{(n)}(\lambda p) \sim \lambda^{4-n}(\ell n\ \lambda)^p \tag{1.7}$$

The more singular behavior (1.5) is thus consistent with (1.7) provided the powers of logarithms in (1.7) sum up to a power. One can therefore say that the singular behavior (1.5) is produced by summing the infrared singularities in the pure Yang-Mills theory. However, according to (B) one can equivalently say that the singular behavior (1.5) stems from the infinitely heavy effective quark mass.

The results (A), (B), and (C) are formulated more precisely in Sections 3 and 4, whereas Section 5 discusses a certain non-analyticity (signifying a phase-transition?) inherent in our approach.

II. ASSUMPTIONS AND PRELIMINARIES

The results mentioned above are based on some assumptions, which are abstracted from perturbation theory. In this section we shall explain these assumptions in some detail.

A. The Decoupling Theorem of Appelquist and Carazzone

The main assumption I am going to make is the abstraction of Appelquist and Carazzone's decoupling theorem.[5] Since this theorem is well-known, I shall not give any kind of derivation. However, for our purpose there are some points which are very crucial. We shall therefore illustrate the decoupling by means of a single example.

The decoupling theorem says essentially that massive fermions decouple in the infrared limit, provided "suitable" renormalization schemes are applied, and provided one chooses a subtraction point μ which is much smaller than the on-shell quark mass m.

If one considers an ultraviolet finite diagram with only external gluons, it is intuitively understandable that massive fermions decouple, since those graphs which involve fermion loops are less singular than those graphs that involve only massless gluons in the infrared limit. Here it is important that the fermion mass is defined in such a way that the inverse quark propagator vanishes for $\not{p}$=m=the quark mass. This definition is in any case possible to any finite order of perturbation theory.

However, the situation is much less clear when we go to ultraviolet infinite graphs, as one can see in a simple way by considering the vacuum polarization tensor,

$$\Pi_{\mu\nu}(k) = (g_{\mu\nu}k^2 - k_\mu k_\nu)\Pi(k). \tag{2.1}$$

To lowest order the fermion contribution to the unrenormalized Π-function is

$$\Pi_o(k) = -\frac{g^2}{2\pi^2} M \int_o^1 d\alpha\ \alpha(1-\alpha)\ln\frac{m^2 - k^2\alpha(1-\alpha)}{\Lambda^2}, \tag{2.2}$$

where M is the number of fermion multiplets. Now, in the Gell-Mann-Low renormalization used by Appelquist and Carazzone the counter term is defined by subtracting at $k^2 = -\mu^2$,

$$\Pi_o(k^2 = -\mu^2) = -\frac{g^2}{2\pi^2} M \int_o^1 d\alpha\ \alpha(1-\alpha)\ln\frac{m^2 + \mu^2\alpha(1-\alpha)}{\Lambda^2}. \tag{2.3}$$

The renormalized Π-function is then obtained by subtracting (2.3) from (2.2), and one gets

$$\Pi_{G-L}(k) = -\frac{g^2}{2\pi^2} M \int_o^1 d\alpha\ \alpha(1-\alpha)\ln\frac{m^2 - k^2\alpha(1-\alpha)}{m^2 + \mu^2\alpha(1-\alpha)}. \tag{2.4}$$

We now see that even when $k^2 \to 0$, this diagram is not necessarily small. However, following Appelquist and Carazzone[5] we now <u>require the subtraction μ to satisfy</u>

$$\mu << m. \tag{2.5}$$

Then it is easily seen by expanding the logarithm in (2.4) that

$$\Pi_{G-L}(k) \sim O\left(\frac{k^2}{m^2}, \frac{\mu^2}{m^2}\right), \tag{2.6}$$

which is arbitrarily small for $k \to 0$ and $\mu << M$.

What happened in the above example is schematically the following:

Unrenormalized behavior	$\ln\frac{\Lambda}{m} + O\left(\frac{k^2}{m^2}\right)$
Counter term for <u>mass-shell</u> renormalization of quark propagator (crucial)	$-\ln\frac{\Lambda}{m} + O\left(\frac{\mu^2}{m^2}\right)$

Behavior of graph with its counter term $\left.\right\} O\left(\frac{k^2}{m^2}, \frac{\mu^2}{m^2}\right) << 1$.

These considerations can be generalized, and I refer to the original paper[5] for all the details.

Since all massive fermions decouple in the above limit, only the pure Yang-Mills diagrams survive. Thus, to any finite order the Greens functions satisfy

$$\Gamma_{G-L}(\lambda p,g,m,\mu,M) \xrightarrow[\substack{\lambda\to 0 \\ \mu/m<<1}]{} \Gamma_{Y-M}(\lambda p,g,\mu) \qquad (2.7)$$

where the subscript's G-L and Y-M indicate the Gell-Mann-Low renormalization scheme and the pure Yang-Mills theory, respectively.

Equation (2.7) means that the infrared behavior of the pure Yang-Mills theory can be constructed from the mass-shell renormalized* quark-gluon theory. This is an important point that we are going to utilize in the following.

So far we have been talking about things that can be shown to be valid in perturbation theory. We are now going to make a step which cannot be shown at present. Namely, we assume that Appelquist and Carazzone's decoupling can be abstracted from perturbation theory, i.e. that Eq. (2.7) is valid in the "exact" (= the sum over all orders) theory. This is the main assumption.

It is clear that when one abstracts a perturbatively correct property this in general enforces new conditions on the parameters entering the theory. Perturbatively the decoupling is valid in the infrared provided $\mu/m<<1$. However, in the abstraction new restrictions may arise. For example, suppose that the Gell-Mann-Low coupling g is very very small. Then, since the infrared pure Yang-Mills theory is expected to be effectively a strong coupling problem, this means that we start out with a renormalized coupling g which is very far from the resulting strong effective coupling. Therefore, it is natural to expect that in order to apply the decoupling theorem non-perturbatively it is not enough to have $\mu/m<<1$, but μ/m should be "extraordinarily" small in order to compensate for the "extraordinarily" small coupling g we start out from. Of course, without knowing the details of the theory, we cannot explicitly write down such a condition. It is, however, important to realize that the abstraction certainly assumes that such new conditions can always

*The coupling constant and the wave function renormalizations are of course performed off mass shell at the point μ.

be realized in a consistent manner.*

The next question is to ask what happens if we do not use a Gell-Mann-Low renormalization scheme. To give an example, let us consider Weinberg's zero-mass renormalization.[3] For finite diagrams there is presumably not much difference (although I am not sure). However, for ultraviolet infinite graphs there is a lot of difference, as we can see by considering again the vacuum-polarization tensor (2.1). In the zero-mass renormalization the counter terms are defined in the massless theory, whereas of course all unrenormalized quantities are the same as before. Thus, instead of the counter term (2.3) we now have

$$\Pi_{o,m=o}(k^2=-\mu^2) = \frac{-g^2}{2\pi^2} M \int_o^1 d\alpha\ \alpha(1-\alpha)\ln\frac{\mu^2\alpha(1-\alpha)}{\Lambda^2} . \qquad (2.8)$$

Subtracting this from Eq. (2.2) we get the Weinberg renormalized Π-function

$$\Pi_W(k) = \frac{-g^2}{2\pi^2} M \int_o^1 d\alpha\ \alpha(1-\alpha)\ln\frac{m^2-k^2\alpha(1-\alpha)}{\mu^2\alpha(1-\alpha)} . \qquad (2.9)$$

We thus see that this renormalization procedure works, since $\Pi_W(k)$ is independent of the cutoff. It was shown in general by Weinberg[3] (by induction) that one always obtains finite results by this procedure. One therefore has[3]

$$\Gamma_W(p,g_W,m_W,\mu) = Z_W(\frac{\Lambda}{\mu}, g_o) \times \Gamma_o(p,g_o,m_o,\Lambda) , \qquad (2.10)$$

where Z_W is independent of the mass (contrary to the Gell-Mann-Low case, where Z_{GL} depends on m/μ), and where

$$m_W = m_o Z_\theta^{-1}(\frac{\Lambda}{\mu}, g_o) . \qquad (2.11)$$

This mass is not what characterizes the mass-shell. Here Z_θ renormalizes the mass-insertion (in the Lagrangian e.g. the term $m_o\bar{\psi}\psi$). The renormalization group constraint can be written

---

*This situation is somewhat analogous to the application of the "old" Gell-Mann-Low renormalization group in the ultraviolet. Here one assumes (consistent with perturbation theory) that when $\lambda \to \infty$ and $\mu >> m$, all masses drop out. If the renormalized coupling g is very large, one probably has to require that μ is "extraordinarily" larger than m, since g = 0 is the appropriate effective ultraviolet coupling.

$$\Gamma_W(\lambda p, g_W, m_W, \mu) = \lambda^{4-n}\Gamma_W(p, g_W(\lambda), m_W(\lambda), \mu) \times \exp -\int_1^\lambda \gamma_W(g(x))\frac{dx}{x} \ , \tag{2.12}$$

where n is the number of external gluons. The effective coupling $g_W(\lambda)$ is defined in the usual way

$$\lambda \frac{d}{d\lambda} g_W(\lambda) = \beta_W(g_W(\lambda)) \ , \tag{2.13}$$

where the β-function is given by

$$\beta_W(g_W) = -\Lambda \frac{\partial}{\partial\Lambda} g_W(\frac{\Lambda}{\mu}, g_o) \ . \tag{2.14}$$

Thus, in the zero-mass renormalization β_W is independent of the mass (whereas β_{GL} is a function of g and m/μ). The effective mass $m_W(\lambda)$ satisfies

$$\lambda \frac{d}{d\lambda} m_W(\lambda) = -[1+\gamma_\theta(g_W(\lambda))]m_W(\lambda), \tag{2.15}$$

where the mass dimension γ_θ is given by

$$\gamma_\theta(g_W) = -\Lambda \frac{\partial}{\partial\Lambda} \ln Z_\theta(\frac{\Lambda}{\mu}, g_o) \ . \tag{2.16}$$

The anomalous dimension is given by

$$\gamma_W(g_W) = -\Lambda \frac{\partial}{\partial\Lambda} \ln Z_W(\frac{\Lambda}{\mu}, g_o) \ . \tag{2.17}$$

The crucial point is thus that β_W, γ_W, and γ_θ do not depend on m_W.

Now let us return to Eq. (2.9) to see how the decoupling works. With $k^2 \to 0$ and $\mu << m$ we find

$$\Pi_W(k) = \frac{-g^2}{2\pi^2} M \int d\alpha \ \alpha(1-\alpha) \ln \frac{m^2 - k^2\alpha(1-\alpha)}{\mu^2\alpha(1-\alpha)} = O(\ln \frac{\mu^2}{m^2}) \text{ for } k \to 0. \tag{2.18}$$

Thus, with $\mu << m$, this graph becomes rather <u>important</u> (almost singular), <u>contrary</u> to what happened in the Gell-Mann-Low case.

Schematically this phenomenon can be represented the following way:

Unrenormalized behavior } $\ln \frac{\Lambda}{m} + O(\frac{k}{m})$

Counter term in zero-mass renormalization	$-\ell n \frac{\Lambda}{\mu}$	No mass - all subtractions done in massless theory at $p^2 = -\mu^2$
Behavior of graph with counter terms	$\ell n \frac{\mu}{m} + O(\frac{k}{m})$	

The conclusion is thus that fermions do not decouple in the infrared in Weinberg's zero-mass renormalization.

It is worth emphasizing that it is not easy to get a simple intuitive feeling for Weinberg's mass m_w. However, the important point is that Weinberg's parametrization of the Greens functions in terms of g_w and m_w provides us with a renormalization framework which works. Whether it is useful or not depends on what one wants to do.

It should also be mentioned that we could equally well have used dimensional regularization, where also the fermions do not decouple.

B. A Non-trivial Infrared Fixed Point

The zero-mass renormalization (or dimensional regularization) can be used in an interesting way if one takes the number of quark multiplets M to be rather large. To see this, consider a theory with Yang-Mills bosons plus M massless fermion multiplets. Let us consider the β-function for this theory,

$$\beta(g) = -\frac{b_o}{16\pi^2} g^3 + \frac{b_1}{(16\pi^2)^2} g^5 + O(g^7). \tag{2.19}$$

As is well known, for the case of color SU(3) one has asymptotic freedom ($b_o > 0$) for $M < 16.5$, whereas one has infrared freedom ($b_o < 0$) for $M \geqslant 16.5$. It was therefore suggested by Gross and Wilczek[6] that by minimizing b_o one could get a non-trivial infrared fixed point. Let us take

$$M = 16.5 - \varepsilon, \quad \varepsilon > 0, \tag{2.20}$$

where ε is e.g. 1/2; then

$$\beta(g_1) = 0 \text{ for } \frac{g_1^2}{16\pi^2} = \frac{b_o}{b_1} << 1, \tag{2.21}$$

since $b_o = O(\varepsilon)$. Computing the two-loop graphs it is found[7] that for M=16

$$\frac{g_1^2}{16\pi^2} = \frac{2}{3}\frac{16.5 - M}{\frac{38}{3}M-102} = \frac{1}{302}\,, \tag{2.22}$$

or, perhaps more appropriately,

$$\frac{g_1^2}{4\pi} \simeq \frac{1}{24}\ (M = 16). \tag{2.23}$$

Thus, this non-trivial fixed point is indeed very small, and hence, one can reasonably argue that perturbation theory can be trusted.

Actually, noticing that in any order of perturbation theory all Greens functions are analytic in the flavor number M (since the Casimirs are polynomials in M), it does make sense mathematically to consider continuous values of M, which means that if we wish we can take ε in (2.20) to be really "epsilonic", $\varepsilon \ll 1$. In this case it is rather reasonable to expect perturbation theory to make sense for the coupling g_1^2 = const. $\times\ \varepsilon$.

<u>Our second assumption consists in assuming that the perturbation zero</u> g_1^2 = const $\times\ \varepsilon$ <u>makes sense in connection with Weinberg's zero mass renormalization</u>.

It is clear that the β-function (2.19) is a Weinberg β-function, since (2.19)-(2.22) were computed in a theory with massless fermions. To see what our assumption implies, let us apply the renormalization group equation (2.12) for M=16.5-ε,

$$\Gamma_w(\lambda p, g_w, m_w, \mu, M) \xrightarrow[M=16.5-\varepsilon]{\lambda\to 0} \Psi(g_w)\lambda^{4-n-\gamma_w(g_1)}\,\Gamma_w\left(p, g_1, \frac{m_w}{\lambda^{1+\gamma_\theta(g_1)}}, \mu, M\right) \tag{2.24}$$

where $\Psi(g_w)$ is a constant (since the zero is of the first order). Here the anomalous dimension $\gamma_w(g_1)$ and the mass dimension $\gamma_\theta(g_1)$ are of course very small, of the order ε.

Equation (2.24) is a standard equation in the case of a non-trivial fixed point. If one expands in powers of g_1, terms of the type

$$g_1^{2n}\ \ell n^q \frac{m_w}{\lambda^{1+\gamma_\theta}\mu} \quad (q \lesssim n), \tag{2.25}$$

dominate. Thus, the small coupling is compensated by an infinite mass, and (2.25) is of the form "0 $\times\ \infty$". Therefore, in spite of the fact that g_1 is very small, one cannot break a perturbation expansion off at any finite order in g_1.

The reason why we need to assume that we can use the fixed point* g_1^2=const.× ε is that the effective coupling for small λ has the form (since the zero of the β-function is of the first order).

$$g(\lambda)^2 = g_1^2 + \text{const.} \times \lambda^p, \tag{2.26}$$

where p is a (small) positive power. In a perturbative manner of speaking, the terms ignored in writing down Eq. (2.24) are therefore of the form

$$\lambda^{pn} \ln^q \frac{m_w}{\lambda^{1+\gamma_{\theta_\mu}}} . \tag{2.27}$$

Thus, the neglected terms are suppressed by powers of λ relative to the leading terms of the type (2.25). It is therefore not unnatural to expect terms of the type (2.27) to be unimportant. Notice that it is essential that the fixed point is of the first order, since for a higher order zero the corrections in (2.26) would only be logarithmically suppressed, and hence there would be no perturbative argument why Eq. (2.24) should be valid.

It should be noticed that the terms we ignore in perturbation theory (2.27) are of a similar nature as those that were ignored in the abstraction of the decoupling theorem. Therefore, writing Eq. (2.24) does not seem more dangerous than abstracting the decoupling. It has, however, been pointed out to me by Symanzik that it may be more dangerous to ignore these terms when one expands in powers of a fixed point coupling, since at the fixed point logarithms sum up to powers. If we take M=16.5-ε with ε a small parameter (which can be varied) this objection does, however, not appear to be so serious, since g_1^2=const.× ε, so that the small correction of order λ^p in (2.26) can be absorbed in ε. In other words, if the correction in Eq. (2.26) is really important, it can be "transformed away" by considering, instead of M=16.5-ε flavors, M'=16.5-ε-const.× λ^p=16.5-ε' flavors! Thus, by changing the theory slightly, one can apparently† get rid of correction terms of the type (2.26) and (2.27). Nevertheless, we prefer to make it an assumption that such terms can be ignored non-perturbatively. We emphasize that this

---

*I thank G. Mack and K. Symanzik for a discussion on this point.

†The point to remember in judging this not quite satisfactory argument is that the Casimir operators in a perturbative expansion of the Greens functions also depend on M, and hence also change (infinitesimally) in going from M to M'.

assumption is correct to any finite order of perturbation theory (in g_1).

III. THE FERMIONIC REPRESENTATION OF THE INFRARED BEHAVIOR OF PURE YANG-MILLS FIELDS

The renormalization schemes of Gell-Mann-Low and Weinberg represent different parametrizations of the same physical Greens functions. Since the perturbative decoupling works in the first scheme but not in the second, it is clear that by abstracting the decoupling from perturbation theory we get non-perturbative constraints of a highly non-trivial character. Since the basic problem with strong coupling theories is to obtain non-perturbative results we may hope that we can gain some interesting results by comparing the different schemes.

We start by noticing that any Greens function in the two schemes must be proportional

$$\Gamma_{G-L}(\lambda p,g,m,\mu,M) = \Phi_\Gamma^{-1}(g, \frac{m}{\mu}, M) \times \Gamma_W(\lambda p,g_W,m_W,\mu,M), \quad (3.1)$$

where we have explicitly introduced the flavor M. By expressing Γ_{G-L} and Γ_W in terms of the unrenormalized Greens functions it is easily seen that

$$\Phi_\Gamma(g,\frac{m}{\mu}) = \left[\frac{Z_3(\frac{\Lambda}{\mu},g_o)_W^{1/2}}{Z_3(\frac{\Lambda}{\mu},\frac{m_o}{\mu},g_o)_{G-L}^{1/2}}\right]^n, \quad (3.2)$$

where Φ_Γ can equally well be expressed as a function of g, m/μ or g_W, m_W/μ. Of course, Φ_Γ cannot depend on Λ/μ, since Γ_{G-L} as well as Γ_W are both finite.

The abstracted decoupling theorem (2.7) can now be applied on the left hand side of Eq. (3.1) and we get

$$\Gamma_{Y-M}(\lambda p,g,\mu) \xrightarrow[\lambda\to 0,\mu/m<<1]{} \Phi_\Gamma^{-1}(g,\frac{m}{\mu},M)\Gamma_W(\lambda p,g_W,m_W,\mu,M). \quad (3.3)$$

Since we have the pure Yang-Mills Greens functions on the left-hand side, all reference to the flavor and the quark mass has dropped out on the left-hand side. Consequently, the same must be true for the right-hand side. In Section 2 we pointed out that to any finite order the quarks do not drop out in Γ_W. The quarks are therefore dynamically active on the right-hand side. We have therefore obtained the result (A) mentioned in the introduction,

$$\begin{Bmatrix}\text{Pure Yang-Mills}\\ \text{theory}\end{Bmatrix} \xrightarrow[\text{distances}]{\text{Large}} \begin{Bmatrix}\text{Yang-Mills plus}\\ \text{fermion representation}\end{Bmatrix}$$

This result was first obtained in ref. 1. Of course, it is important that we have <u>different</u> parametrizations of the <u>same</u> theory on both sides of Eq. (3.3).

It is important to realize that in (3.3) the flavor M can be any number (in particular, we do not have to take M equal to the physical flavor), since the left-hand side ensures a complete degeneracy with respect to flavor. Also, since to any finite order of perturbation theory Γ_W is analytic as a function of M, we need not necessarily take M to be an integer!

With these remarks in mind we can now take $M=16.5-\varepsilon$ [for color SU(3)], in order to apply the non-trivial infrared point discussed in Section 2b. Combining Eq. (3.3) with Eq. (2.24) we thus get

$$\Gamma_{YM}(\lambda p,g,\mu) \xrightarrow[\lambda\to 0,\mu/m<<1,M=16.5-\varepsilon]{} \tilde{\Phi}_\Gamma^{-1}(g,\frac{m}{\mu},M)\lambda^{4-n-\gamma_W(g_1)}$$

$$\times\ \Gamma_W(p,g_1,\frac{m_W}{\lambda^{1+\gamma_\theta(g_1)}},\mu,M), \tag{3.4}$$

where

$$\tilde{\Phi}_\Gamma(g,\frac{m}{\mu},M) = \Phi_\Gamma(g,\frac{m}{\mu},M)\Psi(g,\frac{m}{\mu})^{-1}$$

$$= \Phi_\Gamma(g,\frac{m}{\mu},M)e^{-\int_1^\lambda [\gamma_W(g_1)-\gamma_W(g_W(x))]dx/x} \tag{3.5}$$

where Ψ is the <u>constant</u> factor picked up by applying the renormalization group in Eq. (2.24). Above we used that $g_W = g_W(g,\frac{m}{\mu})$.

It should be noticed that in Eq. (3.4) the effective mass (which is not a "mass-shell mass") is infinite. Thus, although the coupling g_1 is very small, there are singularities coming from the ∞ mass. However, notice that pure Yang-Mills diagrams in an expansion of Γ_W in (3.4) are non-singular. Hence, <u>the only singular diagrams are those that involve at least one fermion loop</u>. We have therefore obtained the result (B) mentioned in Section 1. This result was first obtained in ref. 1.

As discussed in Sect. 2 [see (2.25)] a perturbative expansion of Γ_W in (3.4) involves terms of the type (depending <u>strongly</u> on the flavor)

$$g_1^{2n}\,\ell n^q \frac{m_w}{\lambda^{1+\gamma_{\theta}}\mu} \cdot (q \leq n). \tag{3.6}$$

These terms are of the form "0 × ∞", and consequently perturbation theory cannot be broken off at any finite order. Thus, in spite of the smallness of g_1, we cannot perform practical calculations of the perturbative type. This is, of course, also to be expected since we know that the decoupling can only be valid non-perturbatively in Weinberg's renormalization.

An interesting question is if one can find a physical interpretation of the "fermionic representation" (= the right-hand side of Eq. (3.4)) of the pure Yang-Mills theory? It appears physically interesting that quarks "pop up" even if they were not there in the first place. Of course, one can always dismiss this question by saying that this result is due to the use of a mass (m_w) which is not the "physical" quark mass. Nevertheless, m_w is a parameter which can equally well be used to characterize the physical theory (actually m_w is more like a coupling constant than like a mass), and therefore it does not appear reasonable to dismiss the occurrence of "spontaneously generated quark pairs" as a formal accident.

Perhaps the physical interpretation is that at large distances gauge tubes are formed (a la relativistic vortex lines[8]) and the world viewed from the inside of a gauge tube is locally one-dimensional. Since it is known[9] that in one dimension fermions can be generated from bosons, this may be why one sees that the pure Yang-Mills theory has a fermionic representation. Of course, these remarks are purely speculative, and I do not have any strong evidence that they are right (in other words, they can quite well be wrong!).

IV. POWER BEHAVIOR OF GREENS FUNCTIONS

The question arises whether the fermionic representation of the pure Yang-Mills theory is useful in practice. As pointed out in Section 3, it is certainly not useful to expand in the small coupling. Nevertheless, it can be shown (as was first done in ref. 2) that an interesting result concerning power behavior of the Greens functions can be derived. To see this let us write again Eq. (3.4) in the form

$$\Gamma_{Y-M}(\lambda p, g, \mu) = \tilde{\Phi}_{\Gamma}^{-1}(g_w, \frac{m_w}{\mu}, M)\,\lambda^{4-n-\gamma_w(g_1)}$$

$$\times \Gamma_w(p, g_1, \frac{m_w}{\lambda^{1+\gamma_{\theta}(g_1)}}, \mu, M) + \text{lower order}, \tag{4.1}$$

valid for $\lambda \to 0$ with $\mu/m<<1$ and $M=16.5-\varepsilon$.

Equation (4.1) has the peculiar feature that the left-hand side is independent of the mass scale (m or m_W). The same is therefore true for the right-hand side*. In ref. 2 we deduced from this that Γ_{YM} is power behaved. This was derived by noticing that one trivially has

$$m \frac{\partial}{\partial m} \Gamma_{YM}(\lambda p, g, \mu) = 0 , \tag{4.2}$$

and hence one must also have

$$m \frac{\partial}{\partial m} [\tilde{\Phi}_{\Gamma}^{-1}(g_W, \frac{m_W}{\mu}, M) \times \Gamma_W(p, g_1, \frac{m_W}{\lambda^{1+\gamma_\theta(g_1)}}, \mu, M)]=0, \tag{4.3}$$

if lower order terms in (4.1) are ignored. Taking into account that m_W/m and g_W can be expressed as functions of g and m/μ, Eq. (4.3) leads to a differential equation for Γ_{YM} with the solution (notice that $f_o(g)$ can be a numerical constant)

$$\Gamma_{YM}(\lambda p, g, \mu) \sim \lambda^{4-n-nf_o(g)} . \tag{4.4}$$

The argument of ref. 2 can, however, be somewhat improved so that one avoids differentiating the asymptotic relation (4.1) (which can be a dangerous procedure). In an appendix we have shown that the asymptotic part of Eq. (4.1) can be reduced to solving the functional equation $f(x+y)=f(x)f(y)$, which has an exponential solution (leading to the power behavior (4.4)) for any continuous (but not necessarily differentiable) function.

Comparing the result (4.4) with the perturbative result (1.7) of Poggio and Quinn[4] the question arises whether (4.4) is more or less singular than the perturbative result. This depends on the sign of $f_o(g)$.

The sign of $f_o(g)$ can be obtained by considering Γ_W in Eq. (4.1). From (4.1) and (4.4) we have

$$\Gamma_W(p, g_1, \frac{m_W}{\lambda^{1+\gamma_\theta(g_1)}}, \mu, M) \sim \lambda^{\gamma_W(g_1)-nf_o} . \tag{4.5}$$

If we expand Γ_W in powers of g_1, we have the perturbatively singular quark loop contributions as well as the constant (λ-independent) pure Yang-Mills contributions (proportional to powers of g_1). However, it is easy to see that even non-perturbatively the constant pure Yang-Mills contribution cannot

---

*In the appendix we have explained why the same remark is not very useful if $M=16.5+\varepsilon$ (i.e., for the infrared free case).

dominate, because if it did, Γ_W would be constant. This is only possible from (4.5) if $\gamma_W(g_1)=nf_o(g)$. However, $f_o(g)$ controls the pure Yang-Mills amplitude according to (4.4), and hence $f_o(g)$ is independent of flavor, whereas $\gamma_W(g_1)$ depends strongly on M. Hence, $\gamma_W(g_1)$ cannot be equal to $nf_o(g)$. It therefore follows that since Γ_W certainly contains the constant contributions from the pure Yang-Mills diagrams, and since these contributions cannot dominate $(\gamma_W(g_1)\neq nf_o(g))$, the behavior (4.5) must be singular. Thus

$$nf_o(g) > \gamma_W(g_1). \tag{4.6}$$

In these arguments we have assumed that the pure Yang-Mills graphs contained in Γ_W do not accidentally sum up to zero (this would, if it happened, have to be true for any external momentum p in $\Gamma_W(p)$).

In the Landau gauge (which we used in deriving (4.1)) the sign of $\gamma_W(g_1)$ is positive. It thus follows that

$$f_o(g) > 0\ , \tag{4.7}$$

as a sharp inequality. The <u>result</u> (4.4) <u>is thus more singular than perturbation theory</u>.

In the Landau gauge one finds with M=16 that $\gamma_W(g_1)$ = $25/906 \simeq 0.03$, and hence by means of (4.6) one can sharpen (4.7),

$$f_o(g) > 0.03\ , \tag{4.8}$$

which is probably a rather ridiculous lower limit.

We wish to emphasize that there are two <u>equivalent</u> ways of looking at the singular behavior (4.4), namely

1. The singular behavior arises because perturbative infrared logarithms in the pure Yang-Mills theory sum up to a power.

2. The singular behavior arises from the infinite effective quark mass

$$\frac{m_W}{\lambda^{1+\gamma_\theta}} \xrightarrow[\lambda\to 0]{} \infty\ ,$$

in the fermionic representation of the pure Yang-Mills theory.

V. NON-ANALYTICITY ("PHASE TRANSITION"?)

We would like to end this report with some rather speculative remarks. As emphasized before, one cannot expand the asymptotic functional equation (4.1) in powers of g_1. To give an explicit example, let us consider the propagator function d, for which (4.1) reads (ignoring the constant $\tilde{\Phi}_{\Gamma}$)

$$d_{YM}(\lambda p,g,\mu) \sim \lambda^{2\gamma_w(g_1)} d_W(p,g_1,\frac{m_w}{\lambda^{1+\gamma_\theta(g_1)}},\mu,M). \tag{5.1}$$

Expanding to lowest non-trivial order we get

$$d_{YM}(\lambda p,g,\mu) \sim 1+M\frac{g_1^2}{6\pi^2}\ln\frac{m_w}{\lambda\mu} - \frac{g_1^2}{8\pi^2}(\frac{13}{3}N-\frac{4}{3}M)\ln\lambda + \ldots , \tag{5.2}$$

where the last term comes from the anomalous dimension γ_W in (5.1). Thus, one sees a tendency for M to cancel in the coefficients,

$$d_{YM}(\lambda p,g,\mu) \sim 1 + \frac{11}{2} N \frac{g_1^2}{\pi^2} \ln\frac{m_w}{\mu\lambda^{13/22}} + \ldots , \tag{5.3}$$

where we used that $M = (11/2)N-\varepsilon$ and ignored terms of order ε^2, (g_1^2 is of order ε). Since g_1^2 is strongly M-dependent the terms written down in (5.3) can of course not give a true representation* of d_{YM}.

The conclusion is therefore that if perturbation theory is broken off at <u>any</u> finite order, <u>wrong</u> results are produced, since we cannot get rid of the M dependence in finite orders.

Our relations are thus not analytic in the coupling g_1. Usually, non-analyticity in the coupling signifies phase transitions. It is therefore tempting to speculate that the above non-analyticity is related to a phase transition of some sort. There is, however, apparently no broken symmetry. The speculated "phase transition" would rather be related to a degeneracy in the flavor number M. It would be interesting if one could find some sort of analogy in solid state physics.

ACKNOWLEDGMENTS

I would like to thank N. K. Nielsen and K. Symanzik for their many critical comments and for many discussions on the approach presented above, and H. D. Politzer for much encouragement and for interesting discussions.

---

*It should be noticed that a large N expansion is not possible in the <u>usual</u> sense, since $M=(11/2)N-\varepsilon$, so that <u>both</u> M and N are large. Thus, "leading" as well as "non-leading" terms in N are important. For example, N^2 is of the same order as NM.

I also thank J. Carazzone, G. Mack and E. C. Poggio for their interesting comments and questions.

APPENDIX

In this appendix we shall solve the asymptotic part of the functional equation (4.1) without any mass differentiation. Since we are interested in the λ-dependence we leave out any inessential variables and write Eq. (4.1) in a simplified notation as

$$\Gamma_{YM}(\lambda) = \tilde{\Phi}^{-1}(m_w)\lambda^{4-n-\gamma_w}\,\Gamma_w\left(\frac{m_w}{\lambda^{1+\gamma_\theta}}\right). \tag{A.1}$$

We introduce the function

$$\bar{\Gamma}_{YM}(\lambda^{1+\gamma_\theta}) \equiv \lambda^{n-4+\gamma_w}\,\Gamma_{YM}(\lambda), \tag{A.2}$$

since any function of λ can always trivially be rewritten as a function of $\lambda^{1+\gamma_\theta}$. Then Eq. (A.1) becomes

$$\bar{\Gamma}_{YM}(\lambda^{1+\gamma_\theta}) = \tilde{\Phi}^{-1}(m_w)\Gamma_w\left(\frac{m_w}{\lambda^{1+\gamma_\theta}}\right). \tag{A.3}$$

Introduce the variable

$$\bar{m}_w = m_w/\lambda^{1+\gamma_\theta}, \tag{A.4}$$

Then (A.3) becomes

$$\bar{\Gamma}_{YM}(m_w/\bar{m}_w) = \tilde{\Phi}^{-1}(m_w)\Gamma_w(\bar{m}_w)\ . \tag{A.5}$$

Now take m_w=1 in some arbitrary mass unit. Then

$$\bar{\Gamma}_{YM}(1/\bar{m}_w) = \tilde{\Phi}^{-1}(1)\Gamma_w(\bar{m}_w). \tag{A.6}$$

From this Γ_w can be expressed in terms of $\bar{\Gamma}_{YM}$, and inserting in (A.5) we get

$$\bar{\Gamma}_{YM}(m_w/\bar{m}_w) = \frac{\tilde{\Phi}(1)}{\tilde{\Phi}(m_w)}\,\bar{\Gamma}_{YM}(1/\bar{m}_w)\ . \tag{A.7}$$

In Eq. (A.5) we now put $\bar{m}_w$=1 in some arbitrary mass unit. Then

$$\tilde{\Phi}(m_w) = \Gamma_w(1)/\bar{\Gamma}_{YM}(m_w) = \tilde{\Phi}(1)\bar{\Gamma}_{YM}(1)/\bar{\Gamma}_{YM}(m_w). \tag{A.8}$$

Therefore $\tilde{\Phi}$ can be eliminated in (A.7), and we get from (A.7)

$$\bar{\Gamma}_{YM}(m_w/\bar{m}_w) = \frac{\bar{\Gamma}_{YM}(m_w)}{\bar{\Gamma}_{YM}(1)}\,\bar{\Gamma}_{YM}(1/\bar{m}_w), \tag{A.9}$$

a relation which only involves the single function $\bar{\Gamma}_{YM}$. Introducing a function

$$f_{YM}(\ln z) = \frac{\bar{\Gamma}_{YM}(z)}{\bar{\Gamma}_{YM}(1)}, \tag{A.10}$$

we get

$$f_{YM}(\ln m_w - \ln \bar{m}_w) = f_{YM}(\ln m_w) f_{YM}(-\ln \bar{m}_w), \tag{A.11}$$

i.e. the well known functional equation

$$f_{YM}(x+y) = f_{YM}(x) f_{YM}(y) . \tag{A.12}$$

For any continuous (but not necessarily differentiable) function f_{YM} the unique solution to Eq. (A.12) is*

$$f_{YM}(x) = e^{Cx}, \quad C = \text{constant}. \tag{A.13}$$

From Eq. (A.10) we then get

$$\bar{\Gamma}_{YM}(x) = \bar{\Gamma}_{YM}(1) e^{C \ln x} = \bar{\Gamma}_{YM}(1) x^C. \tag{A.14}$$

From (A.8) and (A.6) one then easily obtains

$$\tilde{\Phi}(x) = \tilde{\Phi}(1) x^{-C}, \tag{A.15}$$

$$\Gamma_w(x) = \tilde{\Phi}(1)\bar{\Gamma}_{YM}(1) x^{-C} . \tag{A.16}$$

Using Eq. (A.2) we get

$$\Gamma_{YM}(\lambda) = \lambda^{4-n-\gamma_w+C(1+\gamma_\theta)}\,\bar{\Gamma}_{YM}(1) . \tag{A.17}$$

---

*I thank K. Symanzik for informing me that differentiability is unnecessary in solving Eq. (A.12). Symanzik also pointed out that non-power behaved <u>discontinuous</u> functions solve Eq. (A.12).

This result has to be independent of M, so if we define

$$f_\Gamma = \gamma_w - C(1+\gamma_\theta), \tag{A.18}$$

f_Γ is independent of M. From Eq. (A.15) we then get

$$-C \ \ell n \frac{m_w}{\mu} = \ell n \ \tilde{\Phi}_\Gamma \ . \tag{A.19}$$

Consequently, from Eq. (A.18) we get an expression for $f_\Gamma(g)$

$$f_\Gamma = \gamma_w(g_1) + [1+\gamma_\theta(g_1)] \frac{\ell n \tilde{\Phi}_\Gamma(g, \frac{m}{\mu}, M)}{\ell n \frac{m_w}{\mu}} \ . \tag{A.20}$$

This result is easily seen to agree with the expression for $f_\Gamma(g)$ in ref. 2 [see Eq. (2.10) in ref. 2] provided the non-perturbative behavior (A.15) is used in the formula of ref. 2. Equation (A.20) is non-perturbative, e.g. since $\tilde{\Phi}_\Gamma$ involves the non-perturbative exponent shown in the last Eq. (3.5).

Finally, we want to mention that if one takes $M=16.5+\varepsilon$, so that one has infrared freedom, one does not get an asymptotic functional equation like (4.1). This is an important point* since intuitively one might feel that it is even better to have a vanishing infrared coupling than to have a small coupling g_1. In the case $M>16.5$, the equation which corresponds to (4.1) reads

$$\Gamma_{YM}(\lambda p, g, \mu) = \bar{\Phi}_\Gamma^{-1}(g_w, \frac{m_w}{\mu}, M)\lambda^{4-n}$$

$$\times \ (\ell n\lambda)^a \Gamma_w(p, g(\lambda)^2 \sim \frac{-1}{\ell n\lambda}, \frac{m_w}{\lambda} (\ell n\lambda)^b, \mu, M)$$

$$+ \text{"lower" orders} \ , \tag{A.21}$$

where a and b are <u>some</u> powers which can be computed, and where $\bar{\Phi}$ differs somewhat from $\tilde{\Phi}$. Introducing a simplifying notation as in Eq. (A.1) we have

$$\Gamma_{YM}(\lambda) = \bar{\Phi}^{-1}(m_w)\lambda^{4-n}(\ell n\lambda)^a \times \Gamma_w(\frac{1}{\ell n\lambda}, \frac{m_w}{\lambda} (\ell n\lambda)^b). \tag{A.22}$$

This equation does not place any restrictions on the λ-behavior. For example, a solution to (A.22) is

---

*I thank G. Mack for stressing this point.

$$\Gamma_w\left(\frac{1}{\ell n\lambda}, \frac{m_w}{\lambda}(\ell n\,\lambda)^b\right)=\left[\frac{m_w}{\lambda}(\ell n\,\lambda)^b\right]^D F(\lambda)$$

$$\bar{\Phi}^{-1}(m_w) = m_w^{-D}$$

$$\Gamma_{YM}(\lambda) = \lambda^{4-n}(\ell n\lambda)^{a+Db}\lambda^{-D}F(\lambda)\ , \tag{A.23}$$

where $F(\lambda)$ is an arbitrary function. Thus $\Gamma_{YM}(\lambda)$ is not at all determined from (A.22).

It is also important to mention in connection with Eq.(A.21) that there is not much reason to expect the lower order terms to be really unimportant. The reason is that the effective coupling has large corrections of the logarithmic type,

$$g(\lambda)^2 \sim \frac{+1}{2b_o\,\ell n\lambda} + 0\left(\frac{\ell n|\ell n\lambda|}{\ell n^2\lambda}\right)\ , \tag{A.24}$$

and hence the "leading" terms in Γ_w in Eq. (A.21) are perturbatively of the type

$$\left(\frac{1}{\ell n\lambda}\right)^n \ell n^q\,\frac{m_w}{\lambda\mu}\ , \quad (q \leqslant n), \tag{A.25}$$

which is at most of order 1. The "non-leading" terms are of the type

$$\left(\frac{1}{\ell n\lambda}\right)^{n-1}\left(\frac{\ell n|\ell n\lambda|}{\ell n^2\lambda}\right)\ell n^q\,\frac{m_w}{\lambda\mu} \quad (q \leqslant n), \tag{A.26}$$

which can be of order

$$\frac{\ell n|\ell n\lambda|}{\ell n\,\lambda}\ , \tag{A.27}$$

which only represents a poor logarithmic suppression relative to the "leading" order one contributions.

From these considerations it is clear that it is essential to choose a non-trivial first order infrared fixed point in order to have a more solid power suppression in non-leading terms [see Eq. (2.27) and compare to Eq. (A.27)].

REFERENCES

1. P. Olesen, Nucl. Phys. B104, 125 (1976).

2. P. Olesen, Niels Bohr Institute preprint, NBI-HE-76-4(1976).

3. S. Weinberg, Phys. Rev. D8, 3497 (1973).

4. E. C. Poggio and H. R. Quinn, Harvard preprint (1976); E. C. Poggio, talk delivered at this workshop.

5. T. Appelquist and J. Carazzone, Phys. Rev. D11, 2856 (1975); K. Symanzik, Comm. Math. Phys. 34, 7 (1973).

6. D. J. Gross and F. Wilczek, Phys. Rev. D8, 3633 (1973).

7. W. E. Caswell, Phys. Rev. Lett. 33, 244 (1974); D. R. T. Jones, Nucl. Phys. B75, 531 (1974); A. A. Belavin and A. A. Migdal, unpublished (1974).

8. H. B. Nielsen and P. Olesen, Nucl. Phys. B61, 45 (1973).

9. T. H. R. Skyrme, Proc. Roy. Soc. A247, 260 (1958); S. Coleman, Phys. Rev. D11, 2088 (1975); A. Luther and V. Emery, Phys. Rev. Lett. 33, 598 (1974).

Currents, Quarks, and Gluons

RABINDRA N. MOHAPATRA*
Department of Physics
City College of C.U.N.Y.
New York, New York

I. INTRODUCTION

The study of the basic constituents of matter and their interactions is a significant part of modern day research in particle physics. The current dogma, built on several years of theoretical and experimental research, is that the basic constituents of hadronic matter are spin 1/2 fermions called quarks[1], and their mutual interactions, be it weak, electromagnetic or strong proceed via the exchange of spin 1 mesons coupling to their sources, which are currents made out of the above fermionic constituents. Further unity of matter and interactions is introduced into this picture by the following twin postulates (which are, of course by no means established by experiments but certainly are not contradicted by them): (i) The equivalence of leptonic and hadronic matter[2] built upon earlier postulates of lepton-quark symmetry[3] and (ii) left-right symmetry[4,5] of all interactions at high energies (or in the symmetric vacuum phase). [In fact, in the symmetric vacuum phase, all symmetries like SU(2), SU(3) etc. are also exact.] The observed asymmetry between leptonic and hadronic matter as well as the non-vector-like character of low energy weak interaction in the current framework[6] of gauge theory are attributed to asymmetries of vacuum. Thus, summarizing our basic set of prejudices, we would like to confront various aspects of these hypotheses with experiments. In particular, we would like to ask:

1. How many constituents of matter are there?
2. Are the spin 1 gluons mediating strong interactions observable?
3. What is the nature of weak currents?

---

*Work supported by National Science Foundation Grant No.MPS 75-07376 and City University of New York Faculty Research Award Programme Grant No. RF-10768

As a prelude to further discussion, we would like to first briefly present the current thinking on the subject and then in Sections 2 and 3 present our new results concerning this question.

A. "Flavor" and "Color"

The quarks are supposed to have the twin attributes, flavor and color[7]. They are independent degrees of freedom. The flavor degree of freedom consists of isospin, hypercharge, charm and maybe more! (All these extra degrees of freedom will be called "charm".) While experiments have established the quantum numbers such as isospin and hypercharge (and of course the associated approximate SU(3) symmetry), charm is yet to be conclusively established. The absence of $\Delta S=1$ neutral currents[8], interpreted within the framework of gauge theories, provides a very strong motivation for at least one "charm" degree of freedom. The existence of four leptons, supplemented by the hypothesis of lepton-quark[3] symmetry also lends additional theoretical support for this idea. Furthermore, if naive parton interpretation of the SPEAR $\sigma(e^+e^- \to \text{hadrons})$ data is correct, that would require several charm quarks. Speculations on this question abound. We have recently presented arguments that the number of flavor degrees of freedom is intimately tied to the number of colors, within a gauge theory framework with physical color[9].

The three quark colors, implicit in the work of Greenberg[7] and explicitly postulated by Han and Nambu[7], are assumed to generate strong interactions through an SU(3) vector local gauge interaction. Within the framework of quark-lepton unification hypothesis[2], the lepton number is treated as an extra color. Lower energy SPEAR e^+e^- data as well as the $\pi^0 \to 2\gamma$ decay width provide indirect evidence for this degree of freedom.

B. Is "Color" Confined or Emitted

At present, there are two points of view concerning the question of whether "color" is an observed or hidden symmetry. One attitude towards "color" is that, the SU(3') local group of strong interactions remains unbroken. In such a picture, since the gauge mesons mediating strong interactions remain massless, and nuclear forces are known to be short range, the "color" gluons must be confined as must all other color non-singlet states!! The true mechanism for this however remains obscure to date. The other alternative, which is the more conventional one, is that SU(3') color gauge mesons are massive. In this case, "color" is an observed symmetry, with "color" singlet states being the lowest lying[11]. Within the gauge theory framework, this gives rise to a non-trivial problem: the reason being that the Higgs-Kibble mechanism, naively implemented, not only breaks the local symmetry but it also completely destroys any trace of the global symmetry from the theory. But, if the model is to be physically acceptable, there must at least be an approximate global "color" SU(3') symmetry left in the theory for classification purposes. We observe that[9] this requirement is very stringent, when we try to implement it within a unified theory framework, and dictates, among other things, that quarks must be integrally charged. This implies that electric charge must receive symmetric contribution from both "flavor" and "color". This, automatically, causes mixing between the weak and strong gauge mesons with profound implications for deep-inelastic lepton hadron scattering[12]. These questions will be discussed in Sec. II and III respectively.

C. Structure of the Weak Current

In the "classical" $SU(2_L)\times U(1)$ gauge theory of Weinberg and Salam[13], the minimal scheme has four quarks and the structure of the charged weak current here is given by,

$$J_\mu = \bar{p}_L\gamma_\mu(n_L\cos\theta + \lambda_L\sin\theta) + \bar{c}_L\gamma_\mu(-n_L\sin\theta + \lambda_L\cos\theta) \qquad (0)$$

The right-handed gauge degrees of freedom remain inert in this model. However,consistent with low energy weak interaction phenomenology, it is possible to have right hand charged currents coupling any of the light quarks n or λ to any of the charmed quarks[14]. This kind of current also has a natural place in left-right symmetric theories[5,15,16]. The two most interesting currents coupling light and heavy quarks are (a) $(\bar{c}n)_R$ type suggested in ref. 14; (b) $(\bar{p}n')_R$ type considered in ref. 5. The $(\bar{c}n)_R$ type current was initially alleged to have difficulty with the K_L-K_S mass difference[5] and nonleptonic weak decays[17]. But we have demonstrated that[16] both the K_L-K_S mass difference calculation of Kingsley _et al_. as well as the non-leptonic decay calculations of ref. 17 are too model dependent to be used to rule out gauge models. Moreover, the large observed asymmetry in $\Sigma^+\to p\gamma$ decay[19] seem to strongly indicate[20] a right-handed current of type $(\bar{c}n)_R$. This current, of course, provides a very natural understanding of the $\Delta I=1/2$ rule observed in non-leptonic decays[21]. These currents will be further analyzed in connection with high energy neutrino scattering in Sec. III.

II. IMPLICATIONS OF UNCONFINED COLOR

In this section we would like to discuss several theoretical aspects of the idea of unconfined "color". As we noted earlier, in the unconfined "color" hypothesis (UCH) the color degree of freedom is treated on the same footing as the flavor degree of freedom. In fact, later on we would like to argue that this similarity is even locally exact. The theoretical points we would like to remind the reader about are:

A. Naturalness of Parity and Strangeness Symmetries

Soon after the advent of gauge theories, it was pointed out by Weinberg [20] that in the framework of gauge theories, where the basic weak coupling (g) is of the same order as the electric charge, the fact that parity and strangeness violation are of order G_F may not be automatic; while at the tree level, the high mass of the weak gauge boson makes the above violations small (G_F) at the level of one loop, the high frequency contributions in general generate parity violating and $\Delta S=1$ corrections of order of the fine structure constant. Weinberg, however, showed in his first paper, that only if strong interactions are generated by an Abelian gluon, the symmetry of the theory allows us to rotate away these order α violations, thus making parity and $\Delta S=1$ effects of order G_F again. Soon afterwards, it was realized by us[23] using different techniques (in the U-gauge[24] and using Bjorken limit procedure) that non-Abelian gauge theories of strong interactions (both with <u>massive</u> and <u>massless</u> gluons) also have the advantage that parity and $\Delta S=1$ effect are naturally of order G_F. The basic ingredient in this proof was the realization that "color" gauge theories of strong interactions involve only spin 1 gluon and the flavor symmetry of the theory (apart from mass breakings) is $U(n)_L \otimes U(n)_R$, n being the number of flavors.

B. Maintaining "Color" as a Global Symmetry

Since the Higgs-Kibble mechanism causes the spontaneously broken generators (or charges) of gauge symmetry to vanish, no trace of the global "color" symmetry is left after the color gluons acquire mass. Thus, it is a non-trivial problem to have massive "color" gluons and yet have a global color symmetry. This problem has been studied by us[25], using the techniques previously developed by Bardakci and Halpern and deWit[26]. The basic strategy is easily summarized for any general gauge group SU(N): Choose N Higgs multiplets $\phi_\alpha^{(i)}$, each belonging to the spinor representation of the gauge group SU(N). Choose the potential in such a way that it is invariant under the bigger group $SU(N) \times \widetilde{SU}(N)$, $\widetilde{SU}(N)$ being an auxilliary symmetry defined over the index i. [The quarks, color gauge bosons, leptons, W-bosons etc., are assumed to be singlets under $\widetilde{SU}(N)$]. It is then possible, by restricting the parameters in the potentials to suitable domains to have

$$\langle\phi_\alpha^{(i)}\rangle = \phi\delta_\alpha^i \ . \tag{1}$$

This breaks all the generators of the local SU(N) but an SU(N)" generated by the sum of the generators of $\widetilde{SU}(N)$ and SU(N), i.e. $F_i''=F_i+\tilde{F}_i$, still remains a good symmetry. Obviously, the quarks and color gluons transform the same way under the surviving group SU(N)" as under the initial local group SU(N). Thus, SU(N)" acts as the actual color symmetry for classification purposes. This theorem has an important corollary which becomes relevant when one tries to apply it to the complex situation where both strong interaction gauge group SU(3') as well as the weak local group G_W are present:

Corollary. If in the U-gauge, a conserved charge Q^i is to survive (like the electric charge), the diagonal elements of $\phi_\alpha^{(i)}$, i.e. $i=\alpha$, must be neutral with respect to Q_i.

Consequences. - 1. Massive Gluons with Fractionally Charged Quarks. In a realistic model, the only conserved charge that remains is the electric charge, Q. Therefore, if Q receives no contribution from color, then the above strategy can always be implemented. However, in this case one always needs an Abelian gauge group U(1) to get the correct electric charge assignment for <u>both</u> quarks and leptons: [Ex. The weak gauge group could be SU(2)×U(1), where U(1) is needed to get correct fractional charge for quarks and it could be, for example, SU(3)×U(1), where U(1) is needed not for quark charges but to get the leptonic charges right.] Thus, in summary, if we allow a U(1) gauge group in the theory, one could have both fractionally charged quarks as well as massive gluons.

In the ultimate theory, since we would like to understand electric charge quantization, we would like to rid ourselves of the U(1) gauge group. This turns out to impose stringent constraints on the theory.

2. Nature of the Local Unifying Symmetry with Massive Gluons. It is now straightforwardly seen that, within the framework of simply unifying groups such as SU(5), E_7, E_6, O(10), SU(16)×SU(16) etc.,[27] requirement of massive gluons and global color symmetry

conflicts with electric charge conservation. To see this, we first observe that the unifying symmetry group $G \supset SU(3') \times G_{flavor}$. To implement our strategy, we need at least 3 identical multiplets transforming as the basic spinor represental of G. The auxilliary symmetry in this case, being a global symmetry, does not contribute to electric charge. Thus, the charge patterns of all three Higgs multiplets are identical. Coming to the case of simple unifying groups proposed in refs.27, at least one multiplet of fermions containing quarks and leptons is assigned to the spinor representation of the group G. Thus, the charge pattern in all three multiplets chosen for our purpose is such that never are <u>all</u> diagonal elements of $\phi_\alpha^{(i)}$ $(i=\alpha)$ electrically neutral. Thus, to the extent that we use only spinor representations of the group to implement our strategy, color gluons cannot be made massive if the unifying gauge group is simple.

On the other hand, if the unifying group is semisimple, it can only be of the form $G_{flavor} \otimes G_{color}$ where both groups are isomorphic to each other. In that case, the gauge group G_{flavor} itself is chosen as the auxilliary symmetry group. We then choose electric charge symmetrically with respect to flavor and "color" as follows,

$$Q = \sum_i a_i [F_{i,color} + F_{i,flavor}] \ . \tag{2}$$

Now, to see how the global symmetry strategy is implemented in this case, we choose a Higgs multiplet, ϕ, to belong to the basic spinor representation (N,N^*) of the gauge group. It is clear from Eq. (2) that diagonal elements of $\phi_\alpha^{(i)}$ are electrically neutral, thereby allowing us by suitable choice of the potential to have,

$$\langle\phi_\alpha^{(i)}\rangle = \phi\delta_\alpha^i \quad \text{for } i=\alpha=1,2,3 \ ,$$

where 1,2 and 3 correspond to the SU(3') subgroup of the color part of the unifying symmetry group.

The implications of massive gluons within unifying gauge theories are therefore, the following:

a) The unifying symmetry has to be of the form $G_{flavor} \otimes G_{color}$.

b) Since the electric charge formula is of the form given in Eq. (2), the quarks must be integrally charged.

c) The number of flavors must be an integral multiple of the number of "colors".

The above discussion provides the basic guidelines in

constructing gauge models, when color gluons are to be massive. We may give a few examples of gauge models where these requirements can be met.

An SU(4)×SU(4') Model[28] and Eight Quark Flavors. This is a purely vector-like model with both right as well as left-handed fermions transforming as the basic spinor representations (4,4*) under SU(4)×SU(4'). The SU(4') contains the quark color group SU(3') with lepton number being the fourth color. To obtain color SU(3') as a global symmetry with massive gluons, one chooses the Higgs multiplet ϕ to transform also as (4,4*) representation of the gauge group. We choose the other Higgs multiplets to be

$$\Sigma: (1,15) \oplus (15,1) \text{ and } \chi: (1,6) \oplus (6,1) .$$

In the presence of the Σ it is possible to choose

$$\langle\phi\rangle = \begin{pmatrix} a & & & \\ & a & & \\ & & a & \\ & & & b \end{pmatrix} \tag{3}$$

[Note that the presence of the Σ with following pattern f is essential for this purpose:

$$\langle\Sigma\rangle = \begin{pmatrix} c & & & \\ & c & & \\ & & c & \\ & & & -3c \end{pmatrix} \tag{4}$$

The equality of the first 3 entries guarantees that "color" symmetry is maintained as a global symmetry.

Vector-like Theory with Non-vectorical Neutral Current. Other important remarks concerning this model are that the Higgs multiplet χ breaks the full group down to a $SU(2)_A \times SU(2)_B \times SU(4')$. Now, it is easy to make the gauge mesons corresponding to $SU(2)_B$ more massive than those corresponding to $SU(2)_A$. As a result, even though the theory was parity conserving (or vector-like) prior to symmetry breakdown, at low energies it will exhibit left-right asymmetry. Then, by a suitable choice of the left and right-handed fermions to the gauge group, the neutral current can be made non-vectorial. To be consistent with low energy phenomenology, one needs in this case eight flavors. This is also a direct consequence of the results stated a few paragraphs back. As a concrete example, we could choose the following assignment of the physical fermions to the gauge group.

$$\begin{array}{l} SU(2)_A \\ \hline SU(2)_B \end{array} \left\{ \begin{pmatrix} p_L \\ n_L(\theta) \\ \hline n'_L(\theta) \\ p'_L \end{pmatrix}_a \begin{pmatrix} c_L \\ \lambda_L(\theta) \\ \hline \lambda'_L(\theta) \\ c'_L \end{pmatrix} \begin{pmatrix} c_R \\ n_R \\ \hline n'_R \\ p_R \end{pmatrix} \begin{pmatrix} c'_R \\ \lambda_R \\ \hline \lambda'_R \\ p'_R \end{pmatrix} \right. . \tag{5}$$

The subscript 'a' denotes the color index and 'a' goes from 1 to 4, (4 denoting the lepton column). So one has also eight leptons in this model, four light, (ν, e^-, μ^-, ν'), and four heavy, (E^o, E^-, M^-, M^o)[29]. It is now clear from the above picture that the neutral current in this theory has mixed chiral structure and is not pure vector. This is certainly in qualitative accord with the present observations of the Gargamelle[30] and HPWF[31] groups.

Models with Six Quark Flavors. Some weak interaction schemes considered recently involve six quark flavors[5,32,33] and six leptons. However, within our scheme, if we accept six quark flavors, we must have six colors. If we leave aside the three quark colors, the remaining three fermionic colors will contain one or two leptonic flavors; (one if all known leptons are assigned to have the same color and two if (ν_e, e^-) and (ν_μ, μ^-) are assigned separate colors). In either case, we have to introduce new kinds of fermions[34]. One may then choose either SU(6)×SU(6') or O(6)×O(6') as the unifying gauge group of nature. In the latter case however, the charge assignment of the constituent hadrons is as follows:

$$\begin{pmatrix} 0 & 1 & 1 & 1 & 1 & 2 \\ -1 & 0 & 0 & 0 & 0 & 1 \\ -1 & 0 & 0 & 0 & 0 & 1 \\ -1 & 0 & 0 & 0 & 0 & 1 \\ -1 & 0 & 0 & 0 & 0 & 1 \\ -2 & -1 & -1 & -1 & -1 & 0 \end{pmatrix} .$$

First of all, this scheme introduces quarks with charges -4/3 (if color is ignored)[34a]. Moreover, one must have either (e^- and μ^+) or (e^+ and μ^-) in one multiplet. However, the nature of strong interaction becomes an intricate question in this model. For example, if the O(3) subgroup of O(6) is used to generate strong interactions, two difficulties arise:

(a) The charge pattern for the p and n color quark becomes

$$\begin{pmatrix} 0 & 1 & 2 \\ -1 & 0 & 1 \end{pmatrix} .$$

This predicts a width for $\pi^0 \to 2\gamma$, which is $9\Gamma(\pi^0 \to 2\gamma)_{expt}$.
(b) One fails to understand why diquark (qq) states do not exist below the three quark baryonic composites.

The other possibility is to generate strong interactions by gauging the SU(3') color subgroup of O(6). Then, it is obvious the leptons will have to have strong interactions. Thus, we would like to suggest that, for models with six flavors and six colors, one ought to use SU(6) ⊗ SU(6') as the unifying group instead of O(6) ⊗ O(6').

III. NEUTRINO INTERACTIONS, QUARKS, CURRENTS AND GLUONS

There have been indications of new phenomena in high energy neutrino[35,36] scattering in both charged current and neutral current interactions: They can be summarized as follows:

(a) Observation of dimuons ($\mu^+\mu^-$) in both neutrino (61 events) and anti-neutrino scattering (7 events). (See Fig. 1 for their x and y distributions.)

(b) Observation of like sign dimuons ($\mu^-\mu^-$) in neutrino scattering.

(c) Rise in $\langle y \rangle^{\bar{\nu}}$ and $\sigma^{\bar{\nu}}/\sigma^{\nu}$ around 50 GeV incident energy.

Now we would like to explore whether there exists any evidence for any of the ideas presented in the previous sections in high energy neutrino scattering. It may be worth pointing out at the start that low energy neutrino and antineutrino data (E<30 GeV) are consistent with the simple spin 1/2 parton model taken in conjunction with the weak current of Eq. (0). The charm changing part of the current of Eq. (0), of course, is not excited at these low energies, if the charm quark mass is around 2 GeV. To fit the details of the low energy data, one seems to require about 6 to 10% antiquarks as part of the sea. [HPWF data gives the fraction of antiquarks $\mathcal{E}\equiv\bar{Q}/Q+\bar{Q}\simeq.06$ from anti-neutrino data for example, and from the fit to the $(d\sigma^{\nu}/dy)$ distribution and also from $\sigma^{\bar{\nu}}/\sigma^{\nu}=.38\pm.02$ at 1 to 11 GeV[37], one gets $\mathcal{E}=.09\pm.04$.]

Coming to the high energy region, the HPWF group reports a rise in $\sigma^{\bar{\nu}}/\sigma^{\nu}$ and $\langle y \rangle^{\bar{\nu}}$ (see the data points of Fig. 1 and Fig.2, respectively). This is of course a clear indication of some new threshold opening up. At present, in our view, there seem to exist two reasonable explanations for this apparent rise. One suggested by Barnett[38] is that the new effect is an indication for the presence of a new righthanded current of type $(\bar{p}n')_R$ with the n' quark mass somewhere between 4 to 5 GeV.

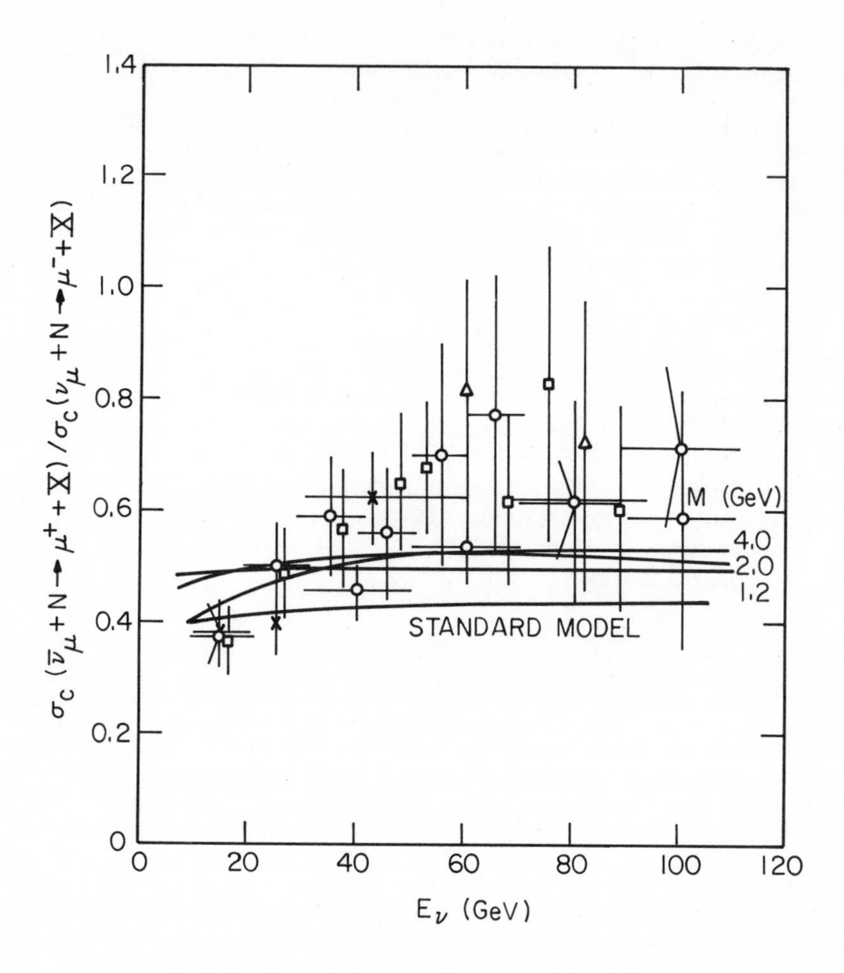

FIGURE 1. Fit to the observed ratio of charged current cross-sections $\sigma^{\bar{\nu}}/\sigma^{\nu}$ in the model with GIM current and with and without massive gluons. Values of the gluon mass shown in the figure. (Ref. 39.)

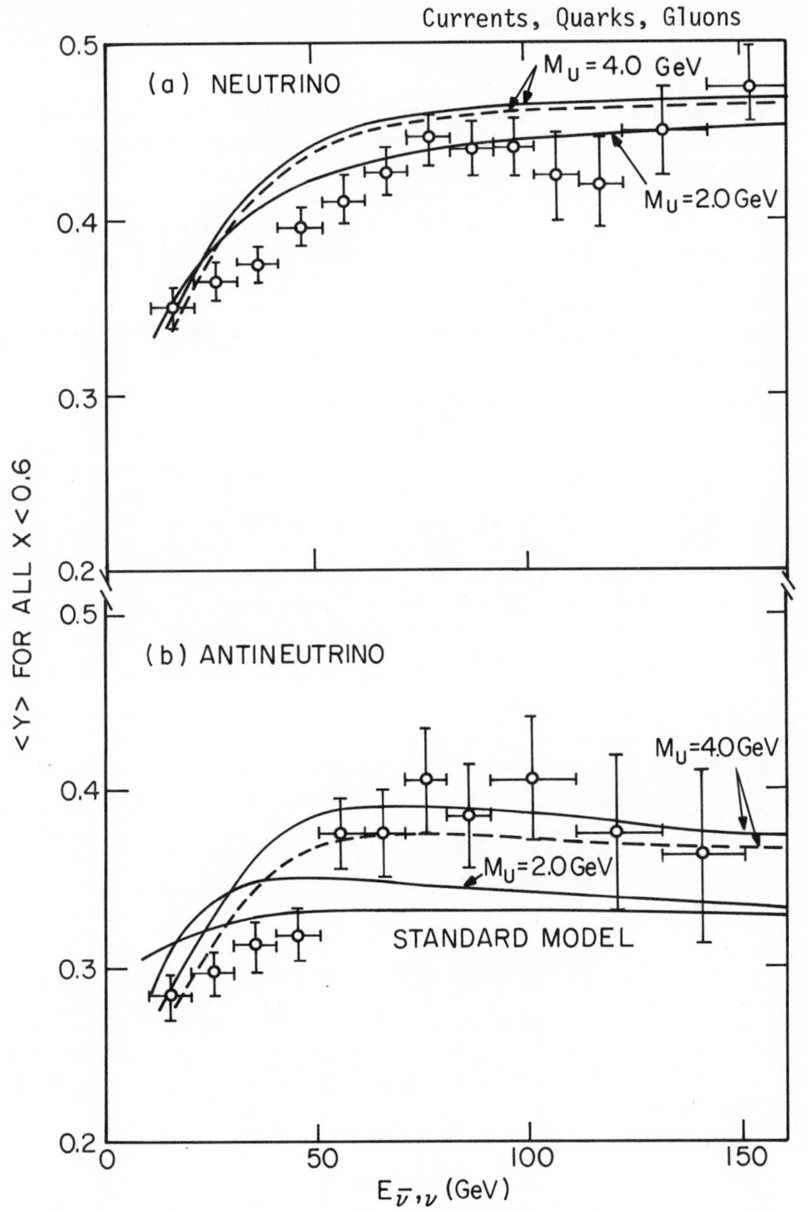

FIGURE 2. Fit to the observed $\langle y\rangle^{\nu}$ and $\langle y\rangle^{\bar{\nu}}$ in the model with GIM current and with and without massive gluons. (Ref. 39.)

The other explanation suggested by Sidhu, Pati and this author[39] is that in theories with massive gluons, one expects color excitation around $E \simeq 50$ GeV both in neutrino and antineutrino scattering. With a color gluon mass $M_{color} \simeq 4$ GeV, we are able to fit both the $\langle y \rangle^{\bar{\nu}}$ and $\sigma^{\bar{\nu}}/\sigma^{\nu}$ data in this model[12,40] without the need for right handed currents of type $(\bar{p}n')_R$. An important point of difference between Barnett's and our work is in the choice of the scaling variable. Since in a model with $(\bar{p}n')_R$ currents one is exciting a heavier quark, one may need to use a different scaling variable,

$$x' = x + \frac{m_{n'}^2}{2Em_N y},$$

where $x=Q^2/2m\nu$. But in our case, except the graphs involving charm changing current $(\bar{c}\lambda)_L$ or $(\bar{c}n)_L$, in all other graphs, one must use the old scaling variable. Since the effect of these graphs is supposed to be small, we use the old scaling variable throughout. To see these points more clearly, we write down the most general weak current for our purposes to be

$$J_{\mu,wk} = \bar{p}_L\gamma_\mu n_L(\theta) + \bar{c}_L\gamma_\mu\lambda_L(\theta) + \sqrt{\alpha_R}\,\bar{c}_R\gamma_\mu n_R + \sqrt{\beta_R}\,\bar{p}_R\gamma_\mu n'_R\,. \tag{6}$$

The differential cross section for neutrinos and antineutrinos off isoscalar targets in the presence of all the factors of Eq. (6) and including the effects of gluon excitation can be written as follows:

$$\frac{d^2\sigma^{(\nu,\bar{\nu})}}{dxdy} = \frac{G_F^2 m_N E}{\pi}\Big[F_2(x,Q^2)(1-y+\frac{y^2}{2}) \mp (y-\frac{y^2}{2})\,xF_3(x,Q^2) + \frac{y^2}{2}(2x\,F_1(x,Q^2)-F_2(x,Q^2))\Big] \tag{7}$$

where F_i are the structure functions and x is the old scaling variable and $y=\nu/E$, where E is the incident energy. Separating F_i into its spin 1/2 part and gluon parts ($F_i=F_{i,flavor}+F_{i,color}$) we obtain (putting $\sin^2\theta=0$)

$$F_{2,flavor}^{(\nu/\bar{\nu})} = \frac{1}{2}\Big[(p(x)+n(x))\begin{Bmatrix}1+\alpha_R\rho_c\\1+\beta_R\rho_{n'}\end{Bmatrix} + (\bar{p}(x)+\bar{n}(x))\begin{Bmatrix}1+\beta_R\rho_{n'}\\1+\alpha_R\rho_c\end{Bmatrix} + 2\begin{Bmatrix}n'(x)\\\bar{n}'(x)\end{Bmatrix}\beta_R\rho_{n'} + 2\begin{Bmatrix}\bar{c}(x)\\c(x)\end{Bmatrix}(1+\alpha_R\rho_c) + 2\begin{Bmatrix}\lambda(x)\\\bar{\lambda}(x)\end{Bmatrix}\rho_c\Big] \tag{8}$$

where the functions $\rho_{c,n'}$ etc. represent the threshold effect, when the c and n' thresholds open. It is chosen in accordance with the approach to scaling of the νW_2 function in deep-inelastic electro-production[40]. p(x), n(x), etc. represent the parton distribution functions in the nucleon[41]. $F_{1,flavor}$ given by the Callan-Gross relation is

$$2\,x\,F_{1,flavor} = F_{2,flavor}\,, \tag{9}$$

and

$$F_{3,flavor}^{\nu/\bar{\nu}}(x) = -\Big[(p(x)+n(x))\begin{Bmatrix}1-\alpha_R\rho_c\\ 1-\beta_R\rho_{n'}\end{Bmatrix} - (\bar{p}(x)+\bar{n}(x))\begin{Bmatrix}1-\beta_R\rho_{n'}\\ 1-\alpha_R\rho_c\end{Bmatrix}$$
$$+2\begin{Bmatrix}n'(x)\\ -\bar{n}'(x)\end{Bmatrix}\beta_R\rho_{n'} + 2\begin{Bmatrix}-\bar{c}(x)\\ c(x)\end{Bmatrix}(1-\alpha_R\rho_c) + 2\begin{Bmatrix}\lambda(x)\\ -\bar{\lambda}(x)\end{Bmatrix}\rho_c\Big]. \tag{10}$$

Now, coming to the gluon part[12], we get,

$$F_{1,color}^{(\overset{\nu}{\bar{\nu}})}(x,Q^2) = \frac{1}{(1+\xi)^2}\left[\frac{1}{2}\sum_{q=p,n,..}(q(x)+\bar{q}(x)) + 8\left(1+\frac{\xi}{4}\right)g(x)\right]\rho_{col.} \tag{11}$$

and

$$F_{2,color}^{(\overset{\nu}{\bar{\nu}})}(x,Q^2) = \frac{x}{(1+\xi)^2}\left[\sum_{q=p,n,..}(q(x)+\bar{q}(x)) + 2\left(3+\xi+\frac{\xi^2}{4}\right)g(x)\right]\rho_{col.} \tag{12}$$

where $\xi=(Q^2/m^2_{color})$ and ρ_{color} is defined in an analogous manner to ρ_c and $\rho_{n'}$. It is clear from Eqs. (11) and (12) that due to the presence of the kinematical factors dependent on ξ, the nonleading contributions are quite dominant for E_ν up to 400 GeV and $E_{\bar{\nu}}$ up to 800 GeV (see Ref. 39 for details). As observed earlier, even for $\beta_R=0$ and $\alpha_R=0$ they provide[39] an adequate description of single muon $\langle y\rangle^{\bar{\nu}}$ and $\sigma^{\bar{\nu}}/\sigma^{\nu}$ (see Fig. 2 and Fig.3).

Ways to Distinguish Between These Two Explanations. (a) To look at σ^ν separately. In Barnett's case, σ^ν is supposed to remain constant at its low energy value except for a small rise due to the excitation of charm quarks due to $(\bar{c}\lambda)_L$ or $(\bar{c}n)_L$ current. On the other hand, in our model, we expect both $\sigma^{\bar{\nu}}$ as well as σ^ν to rise substantially at high energies and the net rise is by equal amounts since gluon contribution is symmetric between neutrinos and antineutrinos. (b) To look at the $(d\sigma/dy)^{\bar{\nu},\nu}$ at very high energies: (E_ν>400 GeV for $m_{color}\simeq$3 GeV). The gluon contribution adds a $(1-y)$ piece to the low energy y-behavior.

Now, coming to the $(\bar{c}n)_R$ type current[14], it is clear from Eqs.(8), (9) and (10), that it has very minimal effect (only due to the charm quark sea) in antineutrino scattering, whereas it adds a new piece to the neutrino scattering, of the form $\alpha_R(1-y)^2(p(x)+n(x))\rho_c$. Thus, taken all together in single muon scattering (both in $\langle y\rangle^{\nu}$ and $\sigma^{\bar{\nu}}/\sigma^{\nu}$) the effect of this current is not too significant. For example, if we ignore the gluon contributions, $\langle y\rangle^{\nu}$ goes from .5 at low energies to $(6+\alpha_R)/(12+4\alpha_R)$ above charm threshold, which for $\alpha_R=1$ is $\simeq .44$. This effect is therefore completely washed away, when gluon contributions are included. Also similar is the case with $\sigma^{\bar{\nu}}/\sigma^{\nu}$.

Dimuons, Gluons and New Currents. We saw in the previous section that the measurement of the neutrino charged current total cross section provides a way to distinguish between the two models which provide adequate description (consistent within present experimental accuracy) of both $\langle y\rangle^{\bar{\nu}}$ and $\sigma^{\bar{\nu}}/\sigma^{\nu}$. Dimuons provide another sensitive test of the gluon vs. $(\bar{p}n')_R$ current idea as well as a test for the presence of $(\bar{c}n)_R$ current. Measurement of the $\langle x\rangle_{\mu\bar{\mu}}$ for neutrinos and antineutrinos provides this crucial test. What the experiments really measure is final visible energy E_{vis}, where

$$E_{vis} = E_H + E_{\mu^-} + E_{\mu^+} . \tag{13}$$

Note that since this does not include the energy of the emitted neutrino in the final state, this is less than the actual incident energy. As a result, $y_{vis} < y$, and since $x=\nu/y$, $x_{vis} > x$. However, they are expected to be close to each other. Now, since the dimuon signal is supposed to result from the decay of either the charmed parton c or n' or the result of the color gluon (as in our model), the $\langle x\rangle$ for the dimuon events is a measure of the $\langle x\rangle$ for their most dominant source. We further know from single muon data that $\langle x\rangle_{valence} \simeq .23$ and $\langle x\rangle_{sea} \simeq .1$. For the gluons, a fit to single muon data indicate[39] (but certainly does not dictate and other choices could also possibly work!) that they are sea-like; therefore $\langle x\rangle_{gluons} \simeq .1$ to .16.

To identify the dominant source for various types of dimuons, we of course need an estimate of leptonic branching ratio for either the charmed particles or vector gluons. For charmed particles (with or without right handed currents and with anomalous dimensions properly taken in account), one estimates[41] the total semileptonic and leptonic branching ratio of about 5 to 10%. With the pure GIM current, then, the dimuon production cross section from valence quarks is due to the presence of $\sin^2\theta$ and since the total sea contribution is bounded by single

TABLE 1. Dimuon Phenomena in Various Models

Nature of Weak Current Responsible for $\mu\mu$ Production	Dominant Source of ν Dimuons	Dominant Source of $\bar{\nu}$ Dimuons	$\langle x\rangle^{\nu}_{\mu\bar{\mu}}$	$\langle x\rangle^{\bar{\nu}}_{\mu\bar{\mu}}$	Comments
GIM $(\bar{c}\lambda)_L(\theta)$	$\sin^2\theta\times$valence + $\cos^2\theta\times$ sea	All sea	$\simeq 0.1$ to 0.15	$\simeq 0.1$ to 0.15	Provides only 10% of observed $\sigma^{\nu}_{\mu\bar{\mu}}$
GIM+$(\bar{c}n)_R$	Valence	Sea	$\simeq 0.25$	$\simeq 0.1$ to 0.15	$\sigma^{\bar{\nu}}_{\mu\bar{\mu}} \simeq (\frac{1}{10}$ to $\frac{1}{5})\sigma^{\nu}_{\mu\mu}$
GIM+$(\bar{p}n')_R$	$\sin^2\theta\times$ valence + $\cos^2\theta\times$ sea	Valence	0.1 to 0.15	0.25	$\sigma^{\nu}_{\mu\bar{\mu}} \simeq (\frac{1}{10}$ to $\frac{1}{5})\sigma^{\bar{\nu}}_{\mu\bar{\mu}}$
GIM+$(\bar{c}n)_R$ + $(\bar{p}n')_R$	Valence	Valence	0.25	0.25	$\sigma^{\nu}_{\mu\bar{\mu}} \simeq \sigma^{\bar{\nu}}_{\mu\bar{\mu}}$
GIM + color gluon	Gluons (sea-like)	Gluons (sea-like)	0.1 to 0.15	0.1 to 0.15	$\sigma^{\nu}_{\mu\bar{\mu}} \simeq \sigma^{\bar{\nu}}_{\mu\bar{\mu}}$
GIM +$(\bar{c}n)_R$ + color gluon	Valence + Gluons	Gluons	$\simeq 0.2$	$\simeq 0.1$ to 0.15	$\sigma^{\nu}_{\mu\bar{\mu}}$ slightly bigger than $\sigma^{\bar{\nu}}_{\mu\bar{\mu}}$

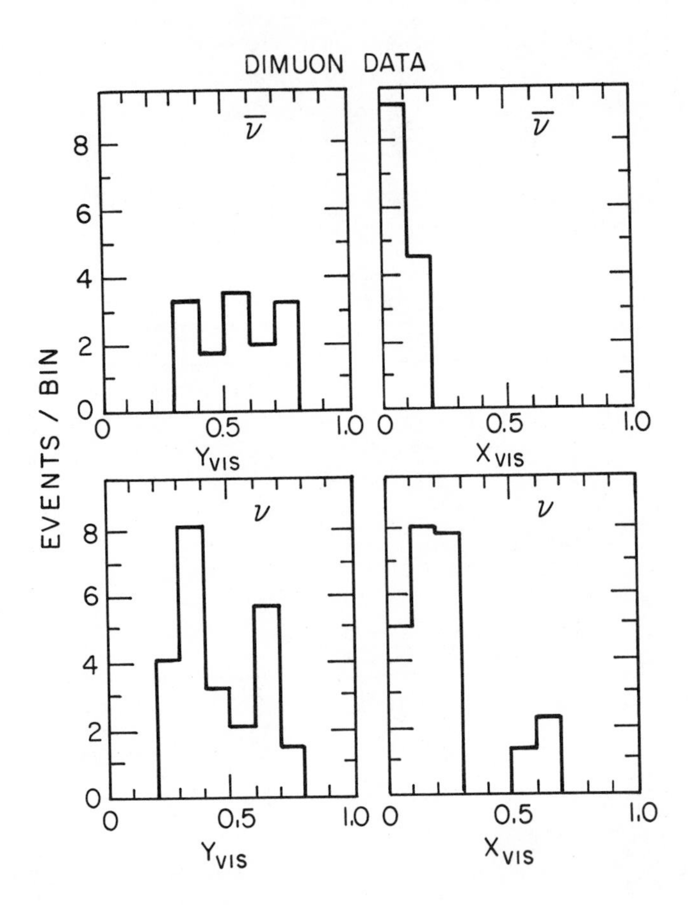

FIGURE 3. Dimuon distribution against x_{vis} and y_{vis}. (From a Review Talk by A.K.Mann, unpublished, 1976). We would like to draw attention to the y distribution for ν-dimuons, which could be of $(1-y)^2$ type indicating possibly $(\bar{c}n)_R$ currents as its dominant source.

muon data to be around 5 to 10%, our view is that pure GIM current provides only about 10% of the observed dimuon cross section. On the other hand the $(\bar{c}n)_R$ current, (since there is no $\sin^2\theta$ suppression) could account for the bulk of the neutrino dimuon cross section. In this case, we predict $\langle x\rangle^{\nu}_{\mu\bar{\mu}} \simeq \langle x\rangle^{\nu}_{\mu}$ and this is roughly in agreement with present observations based on 61 events[35].

IV. CONCLUSION

In these notes, we have tried to argue that, contrary to the prevailing belief, if we assume that color gluons mediating strong interactions are massive (and thus, color is an observable degree of freedom), then interesting restrictions on (i) the type of unifying gauge groups one can use to unify weak, strong, and electromagnetic coupling constants and (ii) the number of flavor degrees of freedom can arise. Of course, whether this picture requires the existence of Higgs scalar particles or dynamical breakdown of color symmetry is at this point purely a matter of speculation. The latter case, however, may provide an easier understanding of why coupling constants split at low energies.

We have then discussed the consequences of this hypothesis for high energy neutrino experiments, for low color thresholds ($M_{color} \simeq 2$ to 6 GeV) and have demonstrated that the rise observed in $\langle y\rangle^{\bar{\nu}}$ and $\sigma^{\bar{\nu}}/\sigma^{\nu}$ in charged current neutrino interactions at high energies can indeed be explained without the need for any new weak currents.

REFERENCES

1. M. Gell-Mann, Phys. Lett. 8, 214 (1964); G. Zweig, CERN Preprint (unpublished) (1964).

2. J. C. Pati and A. Salam, Phys. Rev. D10, 275 (1974).

3. M. Gamba, R. E. Marshak and S. Okubo, Proc. Natl. Acad. of Sci. USA, 45, 881 (1959); J. D. Bjorken and S. Glashow, Phys. Lett., 11, 255 (1964).

4. R. N. Mohapatra and J. C. Pati, Phys. Rev. D11, 566 (1975); ibid, D11, 2558 (1975).

5. Subsequently to the articles of Ref. 4, this idea has also reappeared in literature under the name of vector-like theories (as opposed to the nomenclature of left-right

symmetric theory used in ref. 4). See A. deRujula, H. Georgi and S. L. Glashow, Phys. Rev. D12, 3589 (1975); H. Fritzsch, M. Gell-Mann and P. Minkowski, Phys. Lett. 59, 256 (1976); R. Kingsley, S. B. Treimann, P. Wilczek and A. Zee, Phys. Rev. D12, 2768 (1975).

6. E. S. Abers and B. W. Lee, Phys. Rep. 3C, 1 (1973); M.A.B. Bèg and A. Sirlin, Ann. Rev. of Nuc. Sci. 24, 379 (1974); S. Weinberg, Rev. of Mod. Phys. 46, 255 (1974); J. Illiopoulos, Proc. of the XVII International Conference on High Energy Physics, London (1974).

7. O. W. Greenberg, Phys. Rev. Lett. 13, 598 (1964); M.Y. Han and Y. Nambu, Phys. Rev. 139, B1006 (1965).

8. S. L. Glashow, J. Iliopoulos and L. Maiani, Phys. Rev. D2, 1285 (1970).

9. R. N. Mohapatra and J. C. Pati, "Essential Restriction on the Local Symmetry of a Unified Theory with Massive Gluons", CCNY-HEP-76/7 (1976) (to be published in Phys. Lett. B).

10. J. C. Pati and A. Salam, D8, 1240 (1973); C. Itoh, T. Minamikawa, K. Miura and T. Wabanabe, 1973 (unpublished).

11. O. W. Greenberg and D. Zwanziger, Phys. Rev. 150, 1177 (1966); H. Lipkin, Phys. Lett. B45, 267 (1973).

12. J. C. Pati and A. Salam, Phys. Rev. Lett. 36, 11 (1976); P. Roy and G. Rajasekaran, ibid, 36, 355. For a discussion of this see, J. C. Pati, Proc. of the Northeastern Conf. on "Gauge Theories and Modern Field Theories", edited by P. Nath (1975). See also a related work by G. Furman and G. Komen, Nuc. Phys. B84, 323 (1974).

13. S. Weinberg, Phys. Rev. Lett. 19, 1264 (1967); A. Salam, in Elementary Particle Physics, edited by N. Svarthe (1967) page 367.

14. R. N. Mohapatra, Phys. Rev. D6, 2023 (1972).

15. J. C. Pati and A. Salam, Phys. Lett. 58B, 333 (1975).

16. G. Branco, T. Hagiwara and R. N. Mohapatra, Phys. Rev. D13, 103 (1976).

17. E. Golowich and B. Holstein, Phys. Rev. Lett. 35, 831 (1975).

18. G. Branco and R. N. Mohapatra, Phys. Rev. Lett. 36, 926 (1976).

19. L. K. Gershwin et.al. Phys. Rev. 188, 2077 (1969).

20. M. Ahmed and G. G. Ross, Phys. Lett. 59B, 293 (1975); N. Vasanti, Phys. Rev. D (1976)(to be published).

21. A. deRujula, H. Georgi, and S. L. Glashow, Phys. Rev. Lett. 35, 69 (1975).

21a. For a review of the right-handed current models see, A. deRujula, Invited talk at the Orbes Scientic III,(1976) and also P. Minkowski, same conference, (1976), Cal. Tech. Preprint (1976).

22. S. Weinberg, Phys. Rev. D8, 4482 (1973).

23. R.N.Mohapatra, J.C.Pati and P.Vinciarelli, Phys. Rev. D8, 3652 (1973).

24. R.N.Mohapatra and P.Vinciarelli, Phys. Rev. D8, 481 (1973).

25. R.N.Mohapatra, J.C.Pati and A.Salam, Phys. Rev. D13, 1733 (1976).

26. K.Bardakci and M.B.Halpern, Phys. Rev. D6, 696 (1972).

27. H.Georgi and S.L.Glashow, Phys. Rev. Lett. 32, 438 (1974); F.Gursey and P.Sikivie, Phys. Rev. Lett. 36, 775 (1976); P.Ramond, Cal. Tech. Preprint (1976).

28. R.N.Mohapatra, Phys. Rev. D13, 113 (1976); F.J.Yndurain, CERN Preprint, (1976).

29. For detailed assignment of heavy leptons to the group, see Ref. 28.

30. F. Hassert et.al, Phys. Lett. 46B, 138 (1973).

31. A. Benvenuti et.al. E(11-1) 881-UW-550 (1976).

32. R.M.Barnett, Phys. Rev. Lett. 34, 41 (1975).

33. S. Pakvasa, L. Simmons and S. F. Tuan, Phys. Rev. Lett. 35, 702 (1975).

34. This kind of model will be the subject of a forthcoming investigation by Pati, Salam and the author.

34a. T. Goto and V. S. Mathur, Phys. Lett. B (to be published).

35. A. Benvenuti et.al. Phys. Rev. Lett. 34, 419 (1974); ibid 35, 1199, 1203, 1249 (1975) and Preprints HPWF-76/1 and HPWF-76/2.

36. B. C. Barish et.al. Phys. Rev. Lett. 35, 1316 (1975).

37. H. Deden et.al. Nucl. Phys. B85, 269 (1975).

38. R. M. Barnett, Phys. Rev. Lett. 36, 1163 (1976).

39. D. P. Sidhu, R. N. Mohapatra and J. C. Pati, Brookhaven Preprint (1976).

40. B. C. Barish, Particles and Fields - 1974 (APS/DPF-Williamsburg), ed. by C. E. Carlson.

41. There are various parton distributions available, which are consistent with low energy neutrino and electron scattering data. For a comparison of their relative merits, see the recent paper by J. Okada, S. Pakvasa and D.Parashar, Univ. of Hawaii Preprint (1976). The parton distribution functions with which we have fitted the data of Figs. 2 and 3 are those of R. McElhany and S. F. Tuan, Phys. Rev. D8, 2267 (1973).

42. M. Gaillard, B. W. Lee and J. Rosner, Rev. Mod. Phys. 47, 277 (1974); J. Kandaswamy, J. Schecter and M. Singer, Syracuse Preprint (1976).

Charmonium

TUNG-MOW YAN*†
Laboratory of Nuclear Studies,
Cornell University,
Ithaca, New York

I. INTRODUCTION

It is rather difficult to discuss the structure of hadrons without reference to quarks and partons. Although it is now almost compulsory to identify quarks with partons, they were introduced at different times for different reasons. Quarks were introduced in the low energy regime by Gell-Mann[1] and Zweig[2] to account for the regularities observed in the hadron spectroscopy, while the partons were proposed by Bjorken[3] and Feynman[4] to interpret the scaling phenomenon in high energy inelastic electron scattering. Quarks have the unique characteristic that they carry fractional charges; and the partons are postulated to be structureless. With simple-minded assumptions these ideas have definite experimental consequences.

The low lying hadron spectrum is very well described by a non-relativistic potential quark model. The quark parton model also offers a vivid picture for deep inelastic lepton scattering and e^+e^- annihilation with good results.

Despite the simplicity and success, the dynamical origin of these naive models was not understood when they were suggested. The discovery of asymptotic freedom[5] in non-Abelian color gauge theory has now supplied a theoretical basis for the parton idea and Bjorken scaling. This achievement is made possible by asking the right questions. Instead of the structure functions themselves, one looks for the dependence of the moment of the structure functions on Q^2, the invariant momentum squared of the virtual photon. Bjorken scaling, that the structure function $F_2(x,Q^2)$ should become Q^2 independent as Q^2 approaches infinity, can be restated that its moments are simply constants in this limit:

---

*Supported in part by the National Science Foundation.

†Alfred P. Sloan Foundation Fellow.

$$\lim_{Q^2\to\infty} \int_0^1 dx\, x^{n-2}\, F_2(x,Q^2) = \text{constant} \tag{1}$$

for all $n \geq 2$. By making use of Wilson's operator product expansion[6] and renormalization group equations[7], these moments have very simple structure,

$$\int_0^1 dx\, x^{n-2}\, F_2(x,Q^2) = a_n (Q^2)^{-d_n} \text{ as } Q^2 \to \infty \tag{2}$$

where a_n are constants and d_n are called the "anomalous dimensions" of certain operators. Thus, the validity of Bjorken scaling depends on whether $d_n \equiv 0$ for all n. It has been shown that d_n cannot be all zero in all renormalizable field theories except in the non-Abelian color gauge theories[8]. In these theories d_n are calculable in perturbation by renormalization group technique[5,9]. They are found to behave as[9]

$$d_n = b_n \frac{\ln \ln Q^2}{\ln Q^2}. \tag{3}$$

So d_n tends to zero only slowly as $Q^2 \to \infty$. The moments become

$$\int_0^1 dx\, x^{n-2}\, F_2(x,Q^2) = a_n (\ln Q^2)^{-b_n}. \tag{4}$$

Even the present technology in field theory is not capable of calculating the constants a_n and hence the structure function $F_2(x,Q^2)$; it is still possible to answer the question of Bjorken scaling. In a non-Abelian color gauge theory, Bjorken scaling is violated only logarithmically. The original parton idea, though not perfect, is very close to the theoretical predictions. This example illustrates the importance of formulating the proper questions to be attacked. In this case, the question has been isolated to the vanishing or nonvanishing of the anomalous dimensions of certain operators.

In the quark model for hadron spectroscopy, the important questions are, of course, how quark confinement and correct spectrum arise from a dynamical theory. At present we don't know what is the most efficient way to attack these questions. To me, this is the most important issue in the quark model: to formulate the right problem.

On the phenomenological side, the non-relativistic quark model scores another impressive success in its charmonium model for the newly discovered narrow resonances J/ψ, ψ' and others.

It is hoped that the phenomenological success will generate further stimulus to the efforts directed at the understanding of the dynamical questions of quark confinement and hadron spectrum. Perhaps someday we will understand the naive quark model as we do now the parton model.

II. NAIVE CHARMONIUM

Shortly before the experimental discoveries[10-12] of the J/ψ, Appelquist and Politzer[13] conjectured that the existence of new and heavy charmed quarks c would lead to a family of positronium-like states $c\bar{c}$ lying below the threshold for $e\bar{e} \to$ charmed hadrons[14]. Adopting the current orthodoxy that strong interactions are due to a colored gauge-field, Appelquist and Politzer argued that the large mass m_c of c implied that such systems are small compared to the familiar hadrons, and that, as a consequence, $c\bar{c}$-states would be well described by a Coulombic interaction whose strength--because of asymptotic freedom--would be governed by a relatively small strong fine structure constant α_s. Given this rather weak force, and the heavy mass, they concluded that low lying $c\bar{c}$ states are, to a decent approximation, nonrelativistic. The analogy to positronium would therefore be very close indeed.

An important feature of the Appelquist-Politzer model concerns the decay into ordinary hadrons of $c\bar{c}$ bound states. Such decays proceed via $c\bar{c}$ annihilation into guage gluons--two and three respectively for an even and an odd charge conjugation state (one-gluon annihilation being forbidden because of the color symmetry). It is claimed[13] that a small value of running coupling constant α_s governs these rates, and that the hadronic widths of ^3S $c\bar{c}$ bound states are therefore remarkably small, in accordance with what is observed. (The same argument also relates the width of $\phi \to \rho\pi$ to $\psi \to$ hadrons with considerable success[13].)

Once $\psi' \equiv \psi$ (3684) was discovered[12], it was immediately clear [15] that at least one ingredient of the simple picture had to be modified. In a non-relativistic model, the width of a fermion-antifermion bound state $f\bar{f}$ for decay into $e\bar{e}$ is given by

$$\Gamma(f\bar{f} \to e\bar{e}) = \frac{C}{M^2} |\psi(0)|^2 \qquad (5)$$

where M is the mass of the state, $\psi(0)$ its wavefunction at zero separation, and C a constant. In a Coulomb field, $|\psi_n(0)|^2$ for the $^3S_1$ state with n+1 nodes is proportional to n^{-3}. Assigning $\psi(3684)$ and $\psi(3095)$ to 2^3S_1 and 1^2S_1, respectively, we would expect

$$\left(\frac{3.7}{3.1}\right)^2 \frac{\Gamma(\psi' \to e\bar{e})}{\Gamma(\psi \to e\bar{e})} \simeq \frac{1}{8} \tag{6}$$

whereas the first data gave this ratio as ≃0.7. (Today this ratio is known to be[16] 0.62±0.20.) Now there is a relevant theorem of non-relativistic quantum mechanics: for S-states in a potential V,

$$|\psi(0)|^2 = \frac{m_c}{4\pi} \left\langle \frac{\partial V}{\partial r} \right\rangle . \tag{7}$$

The apparent independence of $|\psi(0)|^2$ on principal quantum number is therefore achieved most simply with a potential that is dominantly linear. This also has the merit that it agrees with the quark confinement forces that gauge theories are surmised to provide[17].

The considerations just sketched lead to what I shall call the naive model[15,18-20]: a non-relativistic pair of quarks, c and $\bar{c}$, moving in a potential that is static to a good approximation. The qualitative features of the spectrum can be surmised[15,18,19] if one recognizes that the potential is less confining than the harmonic oscillator, but more so than the pure Coulomb field. Visual interpolation between these two cases (see Fig. 1) shows that there should be a P-multiplet at 3400-3500 MeV. This immediately brings with it the exciting prediction[15,18,19] of E1 γ-cascades $\psi' \to {}^3P_J \to \psi$. Order-of-magnitude estimates[18] give γ-rates for these transitions in the ∿150 keV range, where they should be easily visible. There should also be pseudoscalar partners (the ${}^1S_0$ state) near and below J/ψ and ψ'. M1 transitions between the ${}^3S_1$ and ${}^1S_0$ states are also to be expected[13,15,18,19].

This rough argument can be augmented by detailed calculations with a specific potential model. In view of the discussion surrounding Eqs. (5) and (6), the most natural choice is[15]

$$V(r) = -\frac{\alpha_s}{r} + \frac{r}{a^2} \tag{8}$$

The parameters α_s, a, and m_c are then obtained by fitting the 2S-1S mass difference to 3684-3095 MeV, and wave functions at the origin to the widths $\Gamma(\psi \to e\bar{e})$ and[21] $\Gamma(\psi' \to e\bar{e})$.

The most important results of these calculations were[15]:

1. The level scheme shown in Fig. 2, with a P-multiplet whose center-of-gravity (c.o.g.) is at ∿3450 MeV, and a D-multiplet at ∿3780 MeV;

2. A total γ-width of ψ' of order 200 keV;

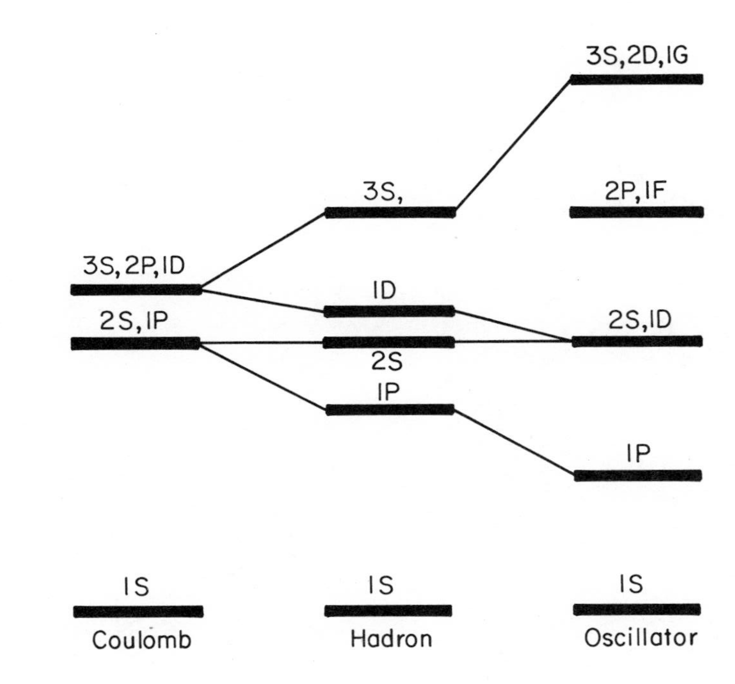

FIGURE 1. The qualitative features of the quark-antiquark excitation spectrum can be inferred by visual interpolation, as shown in this figure. Imagine a continuous deformation of an oscillator potential into a Coulomb field, and the associated change of the bound states. The potential of interest in mesonic spectroscopy is between the two extremes, whence the spectrum labelled "hadron". The parameters of the potentials are chosen so that the 2S-1S difference is fixed. In the case of the "hadron", not all levels are shown for the sake of clarity. Note that we label states by (n+1), where n is the number of radial nodes.

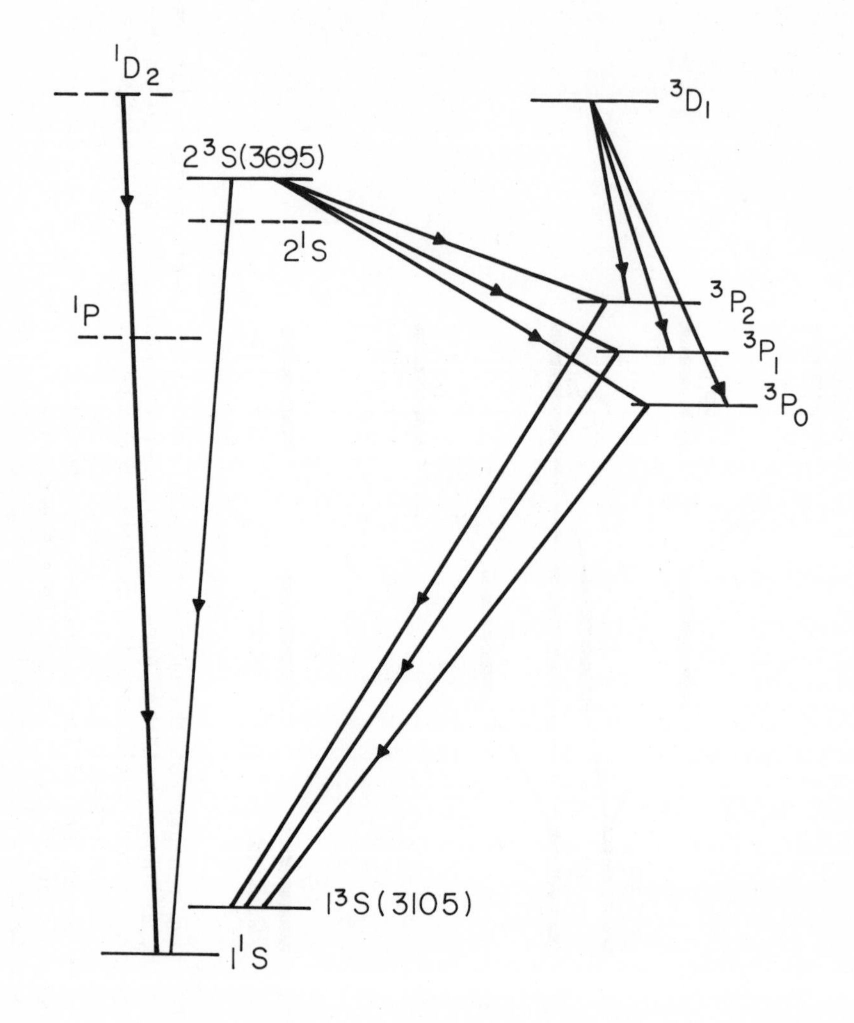

FIGURE 2. The spectrum of charmonium, according to the "naive" model of Ref. 15; the masses of ψ and ψ' are the values of Dec. 1974, not the current values. Comparison with Figs. 1 and 3 is instructive. The heavy lines are allowed E1 transitions, the $2^3S_1 \to 1^1S_0$ transition is a hindered M1.

3. The prediction of a third S-state at ~ 4.15 GeV; with the implication of a further resonance in $e\bar{e} \to$ hadrons;
4. The prediction that the threshold for production of charmed meson pairs should occur at about 3.9 MeV;
5. The observation that the $^3D_1$ state at ~ 3780 MeV could be fed by $e\bar{e}$ annihilation provided there is enough mixing with the nearby $^3S_1$ level (i.e., ψ') to give an appreciable value of $\psi(0)$.

Of equal importance, the parameters that lead to this result are consistent with the ground rules on which the model is based: a small value of α_s ($\sim 1/5$), and non-relativistic motion ($\langle v^2/c^2 \rangle \sim 1/5$).

After many months of disappointing searches[22], γ-transitions were finally found at both[23] DORIS and[24,25] SPEAR. The most essential features of the data are:

1. ψ' decays to four levels called χ at 3.41, 3.45, 3.51 and 3.55 GeV. The three at 3.41, 3.51 and 3.55 are clearly seen in the hadronic decay modes of χ. While the one at 3.45 along with 3.51 and 3.55 are seen in the photon transitions $\chi \to \gamma + J/\psi$, the one at 3.41 is only hinted at in this mode.
2. $\chi(3.41)$ has natural spin parity. The angular distribution of the γ-rays in $\psi' \to \gamma + \chi(3.41)$ is consistent with its being a spin 0 state.
3. The rates for $\psi' \to \gamma\chi$ are far smaller than the original expectation, with a total width[16] of $\psi' \to \gamma$ + anything of order 15 keV at each γ energy.
4. There is a decay[23] $\psi \to \gamma X$, with X at ~ 2.8 GeV.

In addition, a state of 1.865 GeV had recently been seen[26] in e^+e^- annihilation in the energy region between 3.9 and 4.6 GeV. It has the decay modes $K\pi$ and $K3\pi$ expected of a charm meson. This mass is close to the naive estimate of 1.95 GeV.

Chanowitz and Gilman[27] have argued that the most plausible assignment is

$$\begin{aligned}
\chi(3.55)&:\ {}^3P_2,\ J^{PC} = 2^{++}\\
\chi(3.51)&:\ {}^3P_1,\ J^{PC} = 1^{++}\\
\chi(3.45)&:\ {}^1S_0,\ J^{PC} = 0^{-+}\\
\chi(3.41)&:\ {}^3P_0,\ J^{PC} = 0^{++}
\end{aligned}$$

The information of the data plus this assignment is summarized in Fig. 3. A comparison with Fig. 2 reveals, at least at a

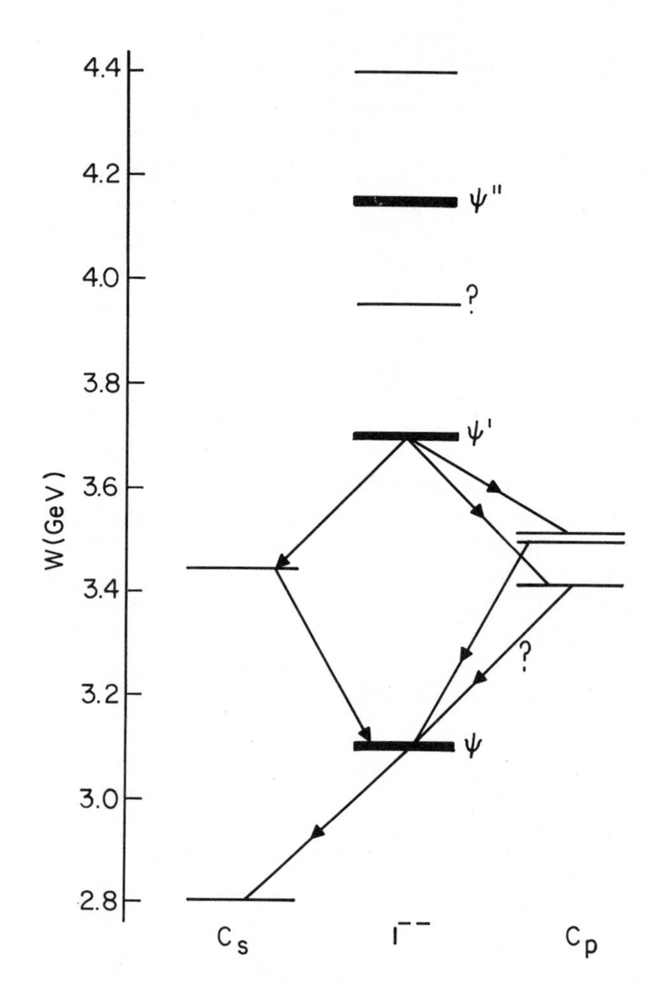

FIGURE 3. The presently known spectrum, and γ-transitions, according to Ref. 10-12, 16,23,24,25,31. The 1^{--} states above ψ' are inferred from R in Fig. 6;whether or not there is a state at $\sim$3.95 GeV is for the reader to judge. The states $C_{P,S}$ are known to have C=+1; S and P refer to their conjectured orbital angular momentum. $C_S(2.8)$ is known[23] to undergo 2γ decay, and $C_P(3.41)$ has even parity, so it cannot be in the C_S column.

superficial level, a remarkable agreement with the predictions.

Leaving the superficial level, let us examine the data more closely. One's first concern may be the small observed γ-rates. Recall, however, that dipole rates are proportional to k^3, and that small errors in level spacings therefore lead to large errors in widths. To illustrate this, assume that $\chi(3.41, 3.51, 3.55)$ are J=0,1,2, respectively. Then recomputing the original[15] widths with the observed k's, one finds $\Gamma(\psi' \to \gamma\ {}^3P_J)$=43,31,37 keV, for J=0,1,2, respectively. The residual discrepancy in widths is therefore a factor of $\sim$3. The more sophisticated model to be described below removes most (though not all) of this discrepancy.

The splittings of the χ-multiplet are considerably larger than had been anticipated on the assumption that only the Coulombic portion of (8) is to be used in computing the Breit-Fermi corrections to the spectrum. It is still not clear whether this represents a fundamental failure of the non-relativistic model, because P-state splittings of order 100 MeV do ensue[28] if the whole of V(r) is treated as the time-component of a 4-potential. Among the three states of χ assigned to the P states, $\chi(3.51)$ decays predominantly to $\gamma+\psi$. Its assignment to ${}^3P_1$ obtains support from the observation[29] that the Landau-Yang theorem forbids such a 1^{++} state to decay into two identical gauge mesons. It must therefore decay into ordinary hadrons by three-gluon annihilation rather than two-gluon annihilation.

There is a possible difficulty for the naive model. If $\chi(3.45)$ were indeed the pseudoscalar partner of ψ', it should have a much larger hadronic width than the other χ's since it is supposed to decay by two-gluon annihilation instead of three-gluon annihilation. Yet no hadronic mode is seen. Furthermore, there is a substantial branching ratio for $\psi' \to \psi + 2\pi$. One would also expect a similar decay for $\chi(3.45) \to X(2.8) + 2\pi$. Again this is not seen. If $\chi(3.45)$ is not the pseudoscalar partner of ψ', what is it?

III. COUPLING TO DECAY CHANNELS[30]

The naive model[15] places the threshold for $e\bar{e} \to$charmed hadrons near 3.9 GeV. The fact that the $e\bar{e}$ cross section shows a sudden rise at about this energy[31], and that the structures at 4.15 and 4.4 GeV have typical hadronic widths in contrast to ψ and ψ' seem to confirm this expectation. Are these the ${}^3S_1$ and ${}^3D_1$ levels expected from the naive model? Furthermore, the electromagnetic decays of ψ' do not agree with the original estimates: the observed strengths[14,15,23,24] for $\psi' \to \gamma\chi$ are several times smaller than predicted, and $\Gamma_e(\psi')=2.1\pm0.3$ keV, in contrast to the theoretical value of 3.0-3.4 keV.

As ψ' appears to be very close to the charm threshold, it may also be unrealistic to describe it as a pure $c\bar{c}$ state; these two problems therefore require a unified treatment.

Aside from these questions specific to charmonium, the coupling to decay channels is a universal problem of hadronic spectroscopy in the quark model. The above questions can be rephrased in more general terms:

1. To what extent does the spectrum of the naive problem survive decay?
2. Do the eigenfunctions of the naive model provide a reliable guide to the internal structure of the states once decay couplings are brought into play?

To investigate these questions we propose a unified model of confinement and decay based on a universal <u>instantaneous</u> interaction between quark color densities ρ_a:

$$H_I = -\frac{1}{2}\int d^3r\, d^3r' \sum_{a=1}^{8} :\rho_a(\vec{r})U(\vec{r}-\vec{r}')\rho_a(\vec{r}'): \quad (\rho_a = \frac{1}{2}\psi^+\lambda_a\psi)$$

Here $U=4r/3a^2$ is the linear[29] potential of the naive model. Hopefully H_I simulates the interaction due to a colored gauge field.

When the quark fields in ρ_a are decomposed into destruction and creation operators, H_I separates into a variety of terms (see Fig. 4). The one in the $c\bar{c}$ sector gives the naive charmonium spectrum, while that in the $c\bar{q}$ sector binds the charmed mesons D, F, etc. There is also a term linking the $c\bar{c}$ and $(c\bar{q},\bar{c}q)$ sectors, which gives the decay amplitudes allowed by the Okubo-Zweig-Iizuka rule. This decay amplitude is schematically given in Fig. 5. Thus the bound states and decay amplitudes are, in principle, determined by a single interaction. Since decay leads to level shifts, the parameters m_c and a have to be readjusted by fitting the ψ-ψ' mass difference, and $\Gamma_e(\psi)$.

All quantities of interest can be extracted from the resolvent $G(z)=P_\psi(z-H)^{-1}P_\psi$, where P_ψ projects onto the $c\bar{c}$ sector. Thus $\mathrm{Im}\langle r=0|G(W+i0)|r=0\rangle$ gives the probability of finding a $c\bar{c}$ pair at zero separation, and hence the contribution δR of charm to R. The poles of G below W_c locate the bound states, and their residues the wavefunctions as modified by virtual decay. From G one can easily compute amplitudes for processes such as $D\bar{D} \to F\bar{F}$ or $e\bar{e} \to D\bar{D}$.

We first evaluate G in the 1^{--} sector by the following steps:

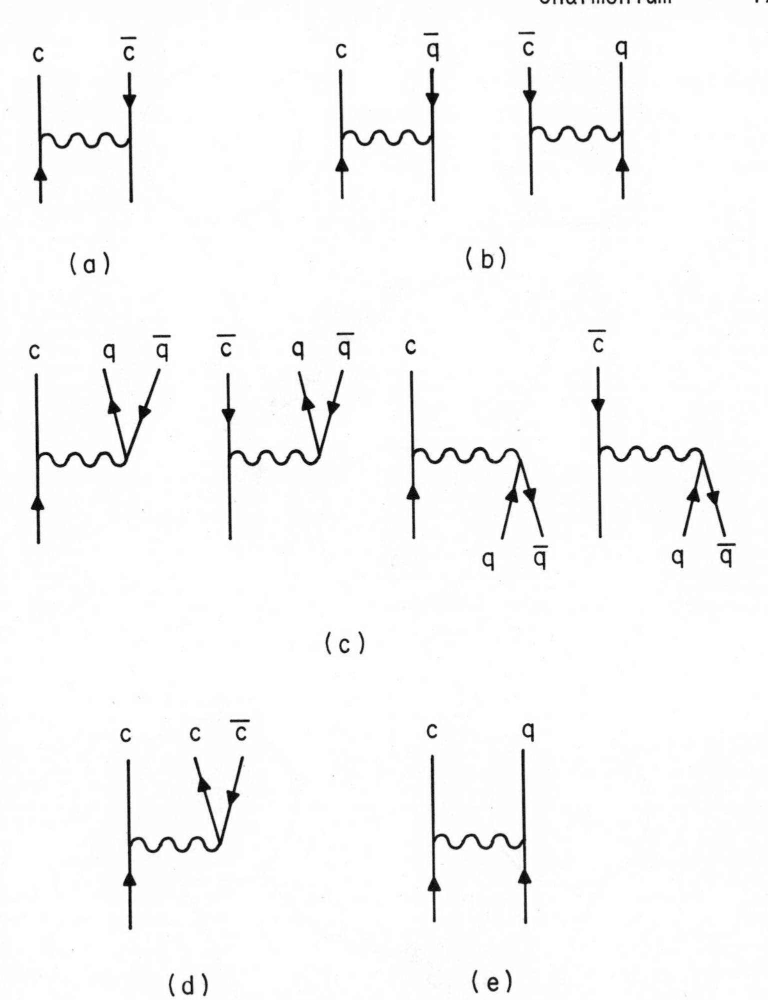

FIGURE 4. Terms in the model's interaction Hamiltonian. The wavy line is the instantaneous potential U. (a) Interactions in the $c\bar{c}$ sector; (b) Interactions that bind charmed mesons; (c) Interactions V that lead to the decays considered; (d) Interactions that go outside the usual framework of the quark model; (e) Terms that contribute to final state interactions.

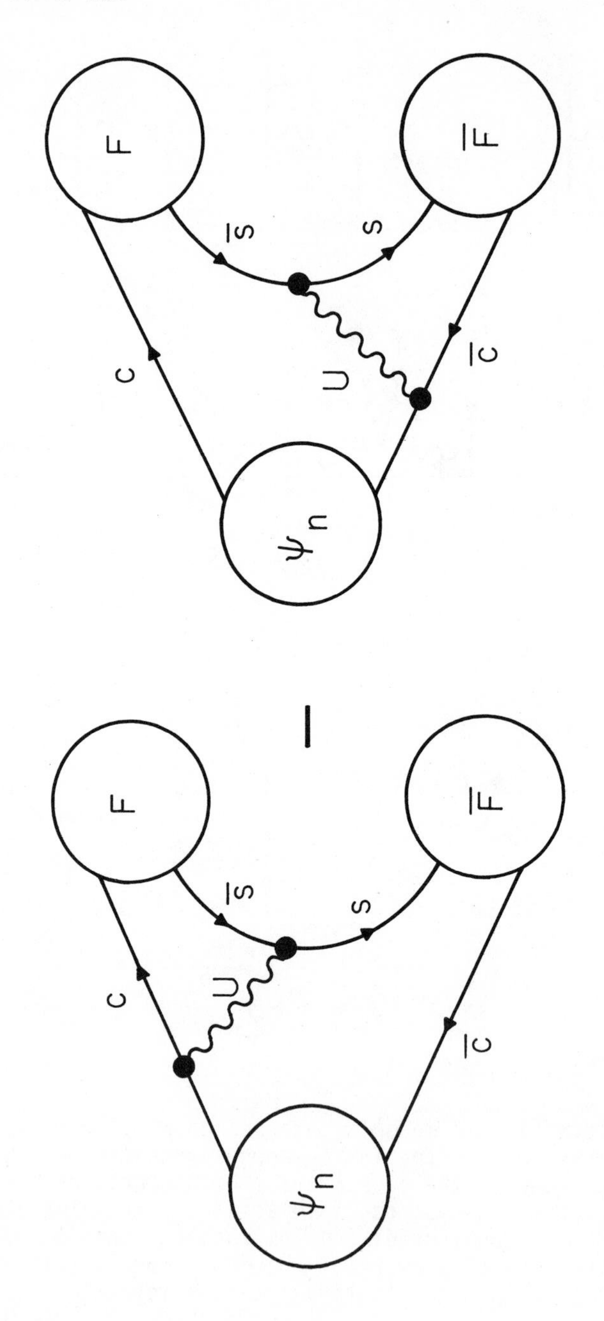

FIGURE 5. The amplitude for decay of an arbitrary $c\bar{c}$ state into a pair of charmed mesons; there are two terms, corresponding to $s\bar{s}$ creation by c or $\bar{c}$.

1. Reduction of terms in H_I to their non-relativistic limits.
2. Evaluation of eigenvalues and eigenfunctions in a $c\bar{c}$ subspace $\mathcal{H}_\psi$ with input parameters a and m_c.
3. Evaluation of the Zweig-allowed decay amplitudes $\mathcal{W}=P_C H_I P_\psi$, where P_C projects onto a subspace $\mathcal{H}_C$ of the $(c\bar{q},\bar{c}q)$ sector; we take $\mathcal{H}_C$ to contain all combinations of D, D*, F, F*.
4. Solution of
$$\mathcal{G} = G_\psi + G_\psi \mathcal{W}^\dagger G_C \mathcal{W} \mathcal{G},$$
where $G_\alpha=(z-P_\alpha H P_\alpha)^{-1}$.
5. Readjustment of m_c and a.
6. Enlargement of $\mathcal{H}_\psi$ until $\mathcal{G}$ stabilizes in the energy regime of interest.

Having solved the 1^{--} sector, we then use the same parameters[32] to evaluate $\mathcal{G}$ in the ^3P_J sectors.

The replacement of a complete set of $(c\bar{q},\bar{c}q)$ states by the subspace $\mathcal{H}_C$ is an increasingly dubious approximation as the energy rises. It is also unlikely that dynamic relativistic effects can be ignored in the decay sector, even if they are only of secondary importance in the $c\bar{c}$ sector. For these reasons our calculations can only be trustworthy for relatively low energies, hopefully $W \lesssim 4.6$ GeV.

IV. RESULTS

I shall now summarize the results[29] of these calculations. The parameters used are given in ref. 32.

Although energy shifts due to virtual decay are quite large, the "renormalized" bound state spectrum is not strikingly different from that of the naive model (see Table I). In particular, the ^3P center of gravity is at 3.44 GeV.

In our decay calculations, explicitly spin-dependent forces (and also the Coulombic term of Eq. (8)) are ignored. On the other hand, we assume mass differences between the 0^- and 1^- charmed mesons[32], and this produces spin splittings of multiplets. In the case of ^3P_J, these are of order 15 MeV, i.e. far smaller than the observed splittings[33], or those obtained from the Breit-Fermi interaction[28].

The ^3P_J and ψ' states are strongly modified by decay. This is demonstrated in Table II, which shows that the probability for finding only a $c\bar{c}$ pair in these states is about 60-80%. These probabilities are comparable for ψ' and the P-states because of two opposing effects: ψ' is closer to the threshold than the P-states, but the latter can undergo S-wave decay whereas ψ' cannot. Further indications concerning the structure of these states can be gleaned from the γ-rates (see especially Table III).

TABLE I. Effects of "Renormalization"

	Bare	"Renormalized" [a]
a	1.0 [b]	1.57
M_c	1.6 [b]	1.89
$M(\psi)$	3.279	3.095
$M(\psi')$	4.053	3.684
$M(\psi'')$	4.704	∿4.175
$M(1^3D_1)$	4.124	∿3.775
$M(2^3D_1)$	4.735	∿4.60
$M(1^3P)$	3.733	3.440

[a] W_c=3.70 GeV. This calculation includes neither Coulombic nor spin-dependent forces.

[b] With Coulomb, but no spin forces (see Ref. 15).

TABLE II. Modification of States Due to Decay[a]

W_c	$^3P_0$	$^3P_1$	$^3P_2$	2^3S_1	$\Gamma_e(\psi')$ [b]
3.70 GeV	0.68	0.75	0.79	0.58	3.6 keV
3.75 GeV	0.71	0.76	0.79	0.67	4.1 keV

[a] The numbers listed under the various states give their norm in the $c\bar{c}$ sector.

[b] These values of $\Gamma_e(\psi')$ would be somewhat lower if Coulomb effects were included.

TABLE III. Radiative Transitions[a]

	E1 Transitions[b,c]		
$\psi' \to {}^3P_J$	J = 0	J = 1	J = 2
$A_{c\bar{c}}$	1.04	1.04	1.04
A_{decay}	0.78	0.39	0.39
A_{pair}	-0.12	-0.05	-0.11
$S \times 10^6$	1.93	1.98	4.46
S/S_o	0.92	0.31	0.42
Γ_J	36 keV	9.7 keV	15 or 2.6 kev[d]

	M1 Transitions[e]		
	$\psi \to \eta_c$	$\psi' \to \eta_c$	$\psi' \to \eta_c'$
Γ	22 keV	10 keV	6 keV

[a] W_c=3.70 GeV. If W_c=3.75 GeV, Γ_0=28 keV, Γ_1=14 keV, and Γ_2=18 or 3 keV.

[b] $A_{c\bar{c}}$, A_{decay}, and A_{pair} are $2^3S_1 \to 2^3P_J$ amplitudes in the $c\bar{c}$ sector, the decay sector, and due to the pair creation term in the current. S is the total strength in GeV^{-2}; it also includes mixing of different S, P and D states. S_0 is the strength in the naive model.

[c] $\Gamma(^3P_J \to \psi\gamma)$=(90,230,280) keV for J=(0,1,2), with $M(^3P_2)$=3.53 GeV.

[d] Γ_2 for $M(^3P_2)$=3.53 or 3.60 GeV, respectively.

[e] Assumed masses for η_c and η_c' are 2.80 and 3.45 GeV.

One would have expected that the depletion of the $c\bar{c}$ component in ψ' would reduce $\Gamma(\psi'\to e\bar{e})$, and improve agreement with the data (2.1±0.3 keV)[16]. Because of subtle interference effects between various $^3S_1$ states that are admixed into $|\psi'\rangle$, the discrepancy is actually increased (see Table II).

The calculation of radiative rates is rather complicated with the new and intricate state vectors. A radiative transition can link the $c\bar{c}$ or the $(c\bar{q},\bar{c}q)$ portions of the participating states. In addition, the electromagnetic current can itself produce a $q\bar{q}$ pair, and thereby cause a transition from the $c\bar{c}$ component of the parent to the $(c\bar{q},\bar{c}q)$ component of the daughter (or vice versa). All these terms have been computed. Fortunately the current's pair production term gives a very small contribution, thereby raising one's confidence in the non-relativistic model being used. The results are summarized in Table III. As we see, some strengths are strongly suppressed (see S/S_0 in Table III) by decay effects. For the E1 transitions $\psi'\to{}^3P_1$ and $\psi'\to{}^3P_2$, these rates are consistent with the fragmentary experimental indications; in the case of $\psi'\to{}^3P_0$, the theoretical rate is too large by a factor of ∿2-3. Here one should remember that even in atomic physics it is difficult to compute E1 rates reliably[34].

In the naive model the M1 transition $\psi'\to\eta_c$ is enormously suppressed[15] because the radial wave functions of $\psi'(\equiv 2^3S_1)$ and the paracharmonium ground state $\eta_c(\equiv 1^1S_0)$ are orthogonal[35]. Because of virtual decay, this orthogonality is destroyed, and $\psi'\to\gamma\eta_c$ has a width (∿2 keV) that may be observable. On the other hand, if η_c is really at 2.8 GeV, the allowed M1 transition $\psi\to\gamma\eta_c$ has a width of over 20 keV, which is somewhat too large. (Once more, the data is extremely sparse.)

R is shown in Fig. 6, together with the preliminary data revealed at the Stanford Conference[31]. As one sees, R has a rich structure above W≃3.9 GeV. Our calculations also show a good deal of structure, though it is not clear whether it is in one-to-one correspondence with the data. The theoretical curve shows a narrow (Γ∿50 MeV) peak at W≃3.75 GeV; this is the 1^3D_1 state. The location, width and area of this peak are very sensitive to the threshold and the spin dependent forces. It is not clear whether this $^3D_1$ state should be identified with the observed structure at ∿3.95 GeV. The prominent theoretical peak at 4.2 GeV is essentially 3^3S_1. The model also has a second sharp (Γ∿50 MeV) $^3D_1$ resonance[36] at W≃4.6 GeV (see Fig. 7). Unfortunately this resonance does not show up in R, even though it is very evident in the many separate terms that go into the calculation of R. For some obscure reason, in this high energy regime our decay amplitudes produce large destructive interference effects that smother this resonance in R. If higher decay

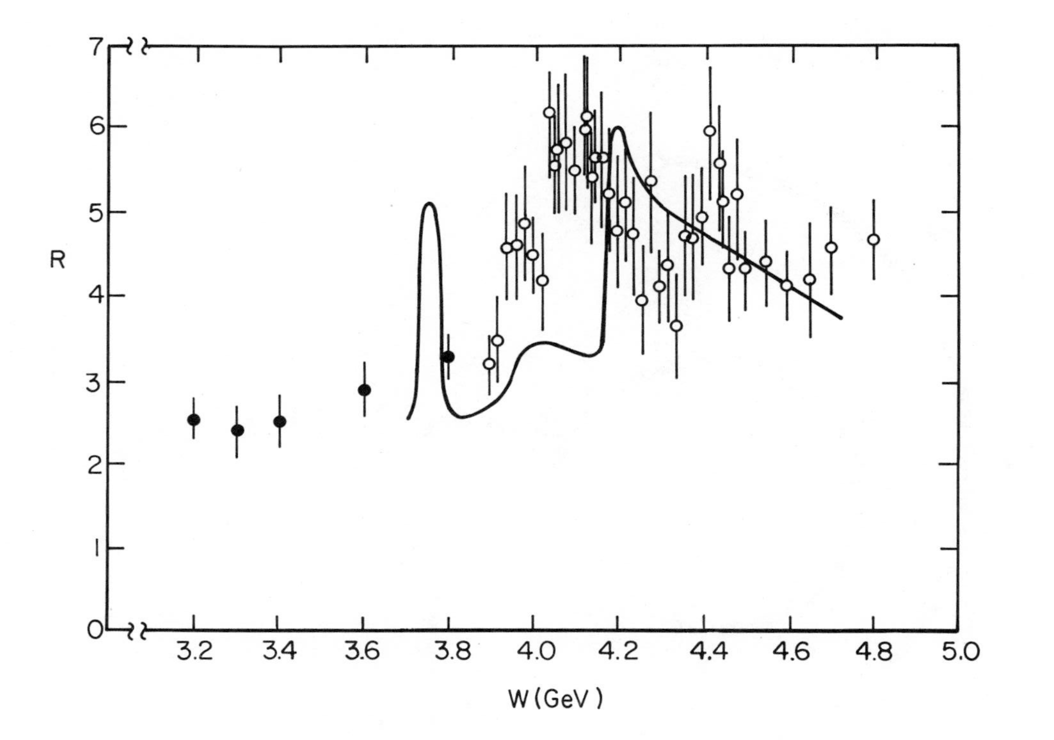

FIGURE 6. R above the charm threshold, taken as W_C=3.70 GeV. A "background" of 2.5 for non-charmed final states is included. The peak at 3.75 GeV is a $^3D_1$ resonance, that at 4.2 GeV is 3^3S_1.

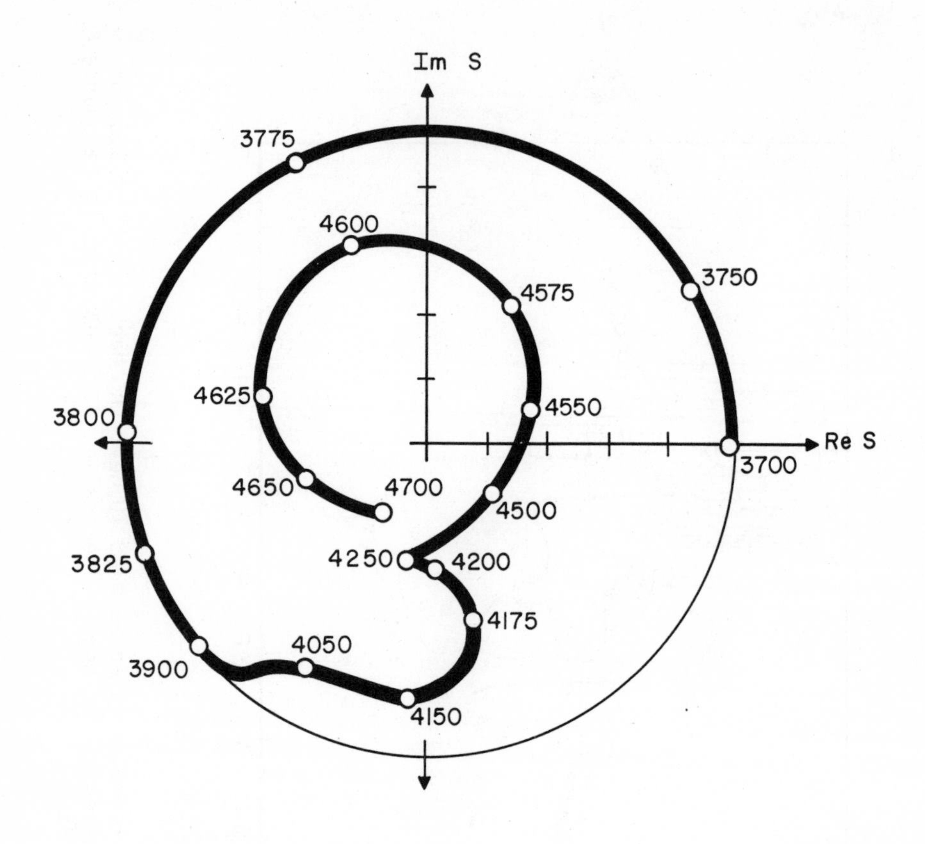

FIGURE 7. The 1^{--} $D\bar{D}$ S-matrix. The resonances at ∿3.775 and ∿4.175 GeV are the $^3D_1$ and $^3S_1$ states of Fig. 6. The inelastic resonance at ∿4.60 GeV is a $^3D_1$ state that does not appear in R because of interference with other levels.

thresholds were to be added (H_C enlarged), as they should be, the ∿4.2 and ∿4.6 resonances would shift downwards; such thresholds might also provide stronger $e\bar{e}$-coupling to the second $^3D_1$-resonance. In any case, it is clear that the observed peak at ∿4.1 GeV is to be identified with 3^3S_1; presumably the observed peak at ∿4.4 GeV is the second $^3D_1$ resonance.

The most favorable reaction[37] for finding charm at an $e\bar{e}$ storage ring is $e\bar{e}\to D\bar{D}$.[38] Our calculations show that except near threshold, this process only provides a small portion of the charm-production cross section. For example, the 4.1 GeV resonance decays mainly ($\gtrsim$2/3) into $D^*\bar{D}^*$, a quite complex final state. If the $1^3\tilde{D}_1$ resonance of Fig. 6 were to exist (present data does not preclude this), it would provide a copious source of slow $D\bar{D}$ pairs. This may still be true of the observed structure at ∿3.95 GeV.

The $^3D_1$ resonance at ∿4.6 GeV mentioned above is seen very strikingly in the Argand plot of $e^{2i\delta}$, where δ is the 1^{--} $D\bar{D}$ phase (see Fig. 7). This plot also shows the first $^3D_1$ resonance, and the 3^3S_1 resonance.

The most striking property of ψ and ψ' are their narrow hadronic width. This is said to be an example of Zweig's rule, though that hardly constitutes an explanation, especially as the Zweig suppression is far more drastic here than in the old spectroscopy. One of the virtues of the Appelquist-Politzer speculation[13] is that this narrow width was forecast as a consequence of the small running coupling constant α_s that pertains to the large masses of the ψ-family. But as we saw in Fig. 8, Zweig-allowed transitions can be combined to form a Zweig-forbidden process, e.g., $\psi \to \phi$. Does this spoil the model by producing a large width for ψ decay into ordinary hadrons? A rough estimate of the ψ-ϕ mixing angle using the decay amplitudes depicted in Fig. 5 leads to a ψ-ϕ mixing angle of ∿10^{-2}; assuming a total effective width of ϕ-like states of 100 Mev, one guesstimates that $\Gamma(\psi\to had)$ is of order 10 keV. Given the crudity of these estimates, this is a very satisfactory outcome[39].

V. CONCLUSIONS

In assessing the results we have described, we note that:

1. Simple ideas work again. The particles in the ψ-family are well represented by a simple charmonium model. By incorporating both confinement and decay, one attains a considerably improved rendition of the data.
2. Fundamental dynamics is largely unknown. Many a particle

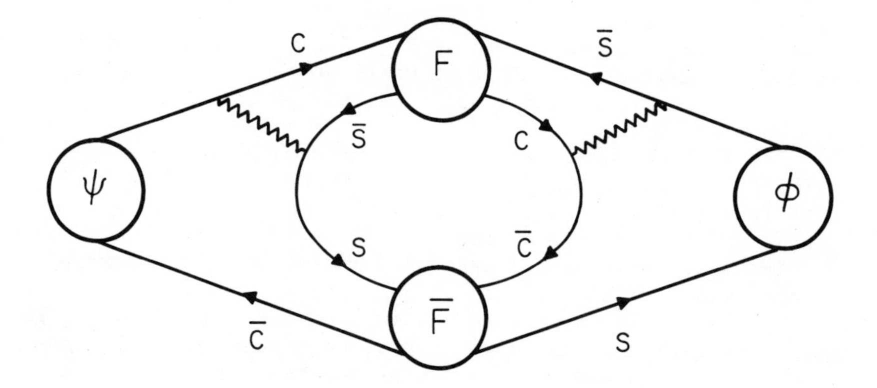

FIGURE 8. A Zweig-forbidden decay $c\bar{c} \to s\bar{s}$ can result from two successive Zweig-allowed decays, as shown here. Nevertheless, the total width for decay of a $c\bar{c}$ state into an $s\bar{s}$ state is estimated to be $\sim$10 keV.

physicist believes that the naive quark model will find its dynamical basis in the non-Abelian color gauge theory. We have made use of some abstraction from such a theory. However, no one has yet been able to demonstrate the connection between the naive quark model and a fundamental dynamical theory.

3. There may be difficulty ahead. The intermediate state $\chi(3.45)$ may turn out to be an embarrassment for the charmonium model. It is also likely that there is more than one new hadronic degree of freedom in the energy regime under study.

REFERENCES

1. M. Gell-Mann, Phys. Lett. 8, 214 (1964).
2. G. Zweig, CERN Reports TH401 and TH412 (1964).
3. J. D. Bjorken, in International School of Physics "Enrico Fermi", edited by J. Steinberger (Academic Press, New York, 1965), Course XLI.
4. R. P. Feynman, Phys. Rev. Lett. 23, 1415 (1969), and in High Energy Collisions (Gordon and Breach, New York, 1969). See also J. D. Bjorken and E. A. Paschos, Phys. Rev. 185, 1975 (1969).
5. H. D. Politzer, Phys. Rev. Lett. 30, 1346 (1973); D. Gross and F. Wilczek, Phys. Rev. Lett. 30, 1343 (1973); G. 't Hooft, unpublished.
6. K. G. Wilson, Phys. Rev. 179, 1499 (1969).
7. C. G. Callan, Phys. Rev. D2, 1541 (1970); K. Symanzik, Comm. Math. Phys. 18, 227 (1970).
8. A. Zee, Phys. Rev. D7, 3630 (1973); S. Coleman and D. Gross, Phys. Rev. Lett. 31, 851 (1973).
9. D. Gross and F. Wilczek, Phys. Rev. D8, 3633 (1973) and Phys. Rev. D9, 980 (1974); H. Georgi and H. D. Politzer, Phys. Rev. D9, 416 (1974).
10. J. J. Aubert et al., Phys. Rev. Lett. 33, 1404 (1974).
11. J.-E. Augustin et al., Phys. Rev. Lett. 33, 1406 (1974).
12. G. S. Abrams et al., Phys. Rev. Lett. 33, 1453 (1974).
13. T. Appelquist and H. D. Politzer, Phys. Rev. Lett. 34,43 (1975). See also A. DeRujula and S. L. Glashow, ibid., 34, 46 (1975).

14. For a review of charm, see M. K. Gaillard, B. W. Lee and J. Rosner, Rev. Mod. Phys. 47, 277 (1975). We use their notation for charmed bosons: D and F are pseudoscalar charmed mesons without and with strangeness, and D* and F* their 1^- counterparts.

15. E. Eichten, K. Gottfried, T. Kinoshita, J. Kogut, K. D. Lane and T.-M. Yan, Phys. Rev. Lett. 34, 369 (1975).

16. V. Lüth et al., Phys. Rev. Lett. 35, 1124 (1975); A. M. Boyarski et al., SLAC-PUB 1599 (1975).

17. S. Weinberg, Phys. Rev. Lett. 31, 494 (1973); A. Casher, J. Kogut, and L. Susskind, ibid., 31, 792 (1973); K. G. Wilson, Phys. Rev. D10, 2445 (1974).

18. T. Appelquist, A. DeRujula, H. D. Politzer and S. L. Glashow, Phys. Rev. Lett. 34, 365 (1975).

19. C. G. Callan, R. L. Kingsley, S. B. Treiman, F. Wilczek, and A. Zee, Phys. Rev. Lett. 34, 52 (1975).

20. See also J. Borenstein and R. Shankar, Phys. Rev. Lett. 34, 619 (1975); B. J. Harrington, S. Y. Park, and A. Yildiz, ibid., 706.

21. In Ref. 15 it proved not to be possible to obtain a completely satisfactory fit to this last piece of data. The closest the model could come to the observed value of 2.1±0.3 keV was 3.0-3.4 keV.

22. J. W. Simpson et al., Phys. Rev. Lett. 35, 699 (1975).

23. W. Braunschweig et al., Phys. Lett. 57B, 407 (1975); B.Wiik, International Symposium on Lepton and Photon Interactions, Stanford, August 1975.

24. G. J. Feldman et al., Phys. Rev. Lett. 35, 821 (1975); W. Tannenbaum et al., ibid., 35, 1323 (1975).

25. G. Goldhaber, invited talk at the Wisconsin International Conference on the Production of Particles with New Quantum Numbers, Madison, Wis., April 22-24, 1976.

26. G. Goldhaber et al., Phys. Rev. Lett. 37, 255 (1976).

27. M. Chanowitz and F. Gilman, SLAC-PUB 1746 and LBL-PUB 4864 (1976).

28. J. Pumplin, W. Repko and A. Sato, Phys. Rev. Lett. 35, 1538 (1975); H. J. Schnitzer, ibid., 35, 1540 (1975).

29. E. Eichten, K. Gottfried, T. Kinoshita, K. D. Lane, and T.-M. Yan, Phys. Rev. Lett. 36, 500 (1976).

30. The literature on decay of ordinary mesons can be traced from A. LeYaouanc, L. Oliver, O. Penne, and J.-C. Raynal, Phys. Rev. D8, 2223 (1973); their treatment has been applied to the ψ-family by R. Barbieri, R. Gatto, R. Kögerler, and Z. Kunszt, CERN TH-2026 (1975). In this approach the mechanism that produces the $q\bar{q}$ pair is not related to the confinement interaction; indeed, the amplitude for $q\bar{q}$ creation is independent of $c\bar{c}$ separation. This is contrary to the picture that $q\bar{q}$ creation results from stretching of the "string" that holds the $c\bar{c}$-pair together. In the treatment of LeYaouanc et al. and Barbieri et al., the W-dependence of decay amplitudes is also ignored--an effect that we find to be important. Another treatment of ψ-family decay, whose objectives and conception are similar to those of Eichten et al.[29], has been given by J. Kogut and L.Susskind, Phys. Rev. Lett. 34, 767 (1975) and Phys. Rev. D12, 2821 (1975). This work is based on the Born-Oppenheimer approximation, which makes detailed comparison with Ref. 29 difficult. The structure of ψ', as described below, is similar to that of Kogut and Susskind, but the spectrum of resonances well above the charm threshold is rather different.

31. R. F. Schwitters, International Symposium on Lepton and Photon Interactions, Stanford, August 1975; J.-E.Augustin et al., Phys. Rev. Lett. 34, 764 (1975).

32. These are m_c, m_q, a, and the masses of the 0^- and 1^- charmed mesons. Our results are obtained with m_c=1.89, m_u=0.31, m_s=0.41, a=1.57. The "bare" 1^{--} masses in $\mathcal{H}_\psi$ are 3.28, 4.05, 4.70 and 5.29 for S-states, and 4.12 and 4.73 for D-states. (All numbers are in GeV or GeV^{-1}.) For $W \lesssim 4.7$, G is insensitive to an enlargement of this $\mathcal{H}_\psi$. The masses in $\mathcal{H}_C$ are M_D=1.85, M_{D*}=2.08, M_F=2.08, and M_{F*}=2.15.

33. It is not ruled out that a relativistic treatment of the light quarks will produce sizeable spin splitting due to virtual decay.

34. For the lighter atoms with one valence electron, single particle model E1 strengths disagree with data by 20-50% [cf. W. L. Miese, M. W. Smith, and B. M. Miles, Nat. Bur. Stand. Ref. Data Ser. 22, (1969)]. In nuclear spectroscopy, the situation is far worse.

35. In this connection, see G. Feinberg and J. Sucher, Phys. Rev. Lett. 35, 1740 (1975), who evaluate relativistic corrections to the $c\bar{c}$ M1 transition amplitude.

36. This resonance is actually an intricate mixture of $^3S_1$ and $^3D_1$ levels, but by studying the calculation in detail, one learns that it originates from 2^3D_1. That the resonances at 3.75 and 4.6 GeV are $^3D_1$-like is shown by the fact that their $e\bar{e}$ width is an order of magnitude smaller than those of ψ,ψ', or the 3^3S_1 resonance at $\sim$4.2 GeV. The data shown in Fig. 6 also shows two resonances with small values of Γ_e.

37. For another proposal, see S. Nussinov, Phys. Rev. Lett. 35, 167 (1975).

38. According to Ref. 26, the state at 1.87 GeV is produced in association with systems of comparable or larger mass. This suggests that it is seen in the channel $e^+e^- \to D\bar{D}^* + \bar{D}D^*$, with D^* being somewhere between 2-2.02 GeV. This suppresses the complications due to cascade decays of D^* such as $D^* \to D\pi$. The channel $e^+e^- \to D\bar{D}$ may already be small at 3.9 GeV.

39. A similar conclusion was reached by a slightly different route by C. Cohen-Tannoudji, C. Gilain, G. Girardi, U. Maor, and A. Morel, Saclay preprint D.Ph.T/75/31 (1975); and by N. A. Törnquist, CERN TH-2002 (1975).

A Phenomenological Model for Charmed Meson Decay

V. S. MATHUR*
Department of Physics and Astronomy
University of Rochester
Rochester, New York

I. INTRODUCTION

Estimates based on quark model show[1] that charmed particles, if they exist, must be quite short-lived with lifetimes typically $\sim 10^{-13}$ sec. Unfortunately one cannot observe directly the tracks of the charge-carrying charmed particle except possibly in emulsion, where such attempts have suffered from poor statistics. The next best thing to do is to look for the weak decay products of the lowest lying charmed particles. The leptonic or semi-leptonic decays, which may have been seen already in dilepton production in neutrino reactions, contain the invisible neutrino, making it hard to deduce the characteristics of the decaying particle. Perhaps the best way of identifying and studying the properties of these particles is through the hadronic decay modes. Indeed, very recently[2] observations have been reported from SLAC of narrow peaks in $K^{\pm}\pi^{\mp}$ and $K^{\pm}\pi^{\mp}\pi^{+}\pi^{-}$ mass spectra at about 1.86 GeV. More peaks in other multihadronic distributions are clearly expected so that a detailed study of the non-leptonic decays of charmed particles acquires considerable importance.

Unfortunately, the non-leptonic decays are not fully understood even for the conventional hadrons. The standard approach is to use group symmetries, current algebra or both. This procedure, however, leads only to sum-rules and does not provide detailed information on the individual decay modes. For charmed particles if we want to know which decay modes are more important or what the individual branching ratios are, it is clear that arguments based on group symmetries or current algebra by themselves are not very useful. Furthermore, since the charmed particles are presumably quite massive, it should be noted that the multihadronic decay modes could be quite important. In such cases, symmetry arguments become particularly intractable. Even

*Research supported by the U.S. Energy Research and Development Administration.

for the simplest two-body decays, in order to correlate decays of charmed mesons into two pseudoscalar, two vector or one pseudoscalar and one vector meson, one would have to employ very high symmetry groups, and the results could be quite unreliable. Alternatively one might use the free quark model to estimate the decay widths. At the high energies available, one might argue on the basis of asymptotic freedom, that one can essentially regard the charm particle decay as the decay of the charmed quark into ordinary quarks, followed by de-excitation into ordinary hadronic matter. Such a procedure, however, would only provide an estimate of the inclusive width, and leave unanswered questions about the relative importance of each exclusive mode of decay.

Besides the hadronic decay, it is important to obtain estimates on leptonic and semileptonic modes of decay. This is of special interest in discussing a possible explanation of the dimuons[3] observed in neutrino and antineutrino induced reactions. Estimates show that in order to understand the rather copious dimuon production on the basis of the GIM model[4], one needs at the very least a rather large branching ratio for the leptonic or semileptonic decay mode.

II. PHENOMENOLOGICAL MODEL

Borchardt, Goto and I have recently studied[5] in detail various decay modes of the charmed mesons on the basis of a simple dynamical model. The model we choose is an obvious generalization of the one proposed by Sakurai[6] to describe the nonleptonic decays of strange particles. To study the parity violating two-body decays of strange particles like $K \to 2\pi$, $\Lambda \to N\pi$, etc. Sakurai suggested that one may employ (i) vector-meson dominance and (ii) universality of the couplings of vector mesons. The vector dominance hypothesis reproduces all the current algebra constraints on $K \to 2\pi$ and the K* dominance in parity violating hyperon decays leads to a reasonable description if we take the K* couplings to baryon-antibaryon pairs to be SU(3) invariant. Furthermore, if we assume that K* is coupled universally to the appropriate strangeness changing component of F-spin in the same way that ρ is coupled universally to the iso-spin, the couplings of K* to $K\pi$ and to baryon-antibaryon pairs are related, so that the parity violating hyperon decay amplitudes can be calculated from the $K_1^0 \to \pi^+\pi^-$ decay matrix element, and the results are in rough agreement with the experimental values.

In order to see how the model is related to the standard theory, consider the weak Hamiltonian in the current-current form

$$H_W = \frac{G_F}{\sqrt{2}} j_\mu j_\mu^+ + h.c. \ , \tag{1}$$

where $j_\mu = \ell_\mu + J_\mu$, ℓ_μ and J_μ being the usual leptonic and hadronic weak V-A currents. In the Cabibbo form

$$J_\mu = \cos\theta_c (V_{\mu 1}^2 + A_{\mu 1}^2) + \sin\theta_c (V_{\mu 1}^3 + A_{\mu 1}^3) \ , \tag{2}$$

where $V_{\mu\beta}^\alpha$ and $A_{\mu\beta}^\alpha$ ($\alpha,\beta=1,2,3$) are the octets of vector and axial-vector currents.[6] For the non-leptonic decays, if we assume that the octet is dynamically enhanced, the non-leptonic piece of the Hamiltonian may be written as

$$H_W^{N.L.} = x \frac{G_F}{\sqrt{2}} (J_\mu J_\mu^+)_{8_s} + h.c. \ , \tag{3}$$

where x represents the enhancement factor, which can be calculated in gauge theory models if the strong interactions are taken to be asymptotically free. However, if we are only interested in the ratios of decays, we do not need the explicit value of x.

Now, the model we use is implemented by the following phenomenologically motivated relations between currents and fields

$$V_{\mu\beta}^\alpha = \sqrt{2} \ \frac{m_v^2}{f_v} \phi_{\mu\beta}^\alpha$$

$$A_{\mu\beta}^\alpha = \sqrt{2} \ f_p \ \partial_\mu P_\beta^\alpha \tag{4}$$

where $\phi_{\mu\beta}^\alpha$ and P_β^α are the octets of vector and pseudoscalar meson fields and f_V and f_P represent the appropriate coupling constants. The non-leptonic decays are then effectively described by a two-meson vertex. The vector coupling constant for the ρ-meson can be estimated from the experimental rate[7] for $\rho \to \ell\bar{\ell}$ decay, which yields

$$\frac{f_\rho^2}{4\pi} = 2.12 \ {}^{+.28}_{-.22} \ . \tag{5}$$

The coupling constant for K* can be estimated using the (first) spectral function sum-rule[8] for SU(3), which on saturation by vector mesons alone, gives $m_\rho^2/f_\rho^2 = m_{K*}^2/f_{K*}^2$, so that

$$\frac{f_{K*}^2}{4\pi} \simeq 2.85 \ . \tag{6}$$

We ignore the contribution of the scalar meson κ, whose existence is not certain. Note also that the coupling of the scalar κ to the strangeness changing vector current vanishes in the SU(3) limit. The pseudoscalar coupling constant for the pion can be determined from $\pi \to \ell \bar{\nu}_\ell$ decay, and assuming SU(3) invariance we take

$$f_K \simeq f_\pi \simeq 92.6 \text{ MeV} . \tag{7}$$

As mentioned before, the parity violating decays are well described by the model. The parity conserving weak interaction represented by two pseudoscalar (or two vector) meson vertices can be tested in p-wave hyperon decays. The pseudoscalar meson poles in this case (there are no vector meson poles for parity conserving hyperon decays) make a negligible contribution, however. This situation should be compared with the current algebra calculation for parity conserving hyperon decays where the current commutator term makes a vanishing contribution (in the SU(3) limit). It is well known[9] that the baryon pole terms are important in this case and both the current algebra approach and the present model should presumably be supplemented by these terms. However, in the present work we will confine ourselves only to meson decays, so these problems do not appear.

The $K_1^0 \to \pi^+\pi^-$ decay proceeds in the model through the K* pole. The strong coupling constant for the $K^*\bar{K}\pi$ vertex can be estimated from the K* width. With f_{K^*} given by Eq. (6), $\tan\theta_C \simeq 0.25$, $G_F \sim 10^{-5}/M_p^2$ and knowing the experimental value of the $K_1^0 \to \pi^+\pi^-$ decay rate[7], we estimate

$$x \simeq 9.1. \tag{8}$$

This value of the enhancement factor is somewhat larger than the recent theoretical estimates[10] which, however, are too low to fully account for octet enhancement.

The present model also successfully describes the leptonic and semi-leptonic decay of ordinary mesons. For the pure leptonic decay, there is no difference from the standard treatment. For the semi-leptonic decay like $K_{\ell 3}$, the hadronic matrix element is usually written in terms of the $f_\pm$ form-factors. If however, the leptons are assumed to be massless, the f_- form-factor does not contribute. Furthermore, if f_+ is assumed to satisfy an unsubtracted dispersion relation dominated by the K* pole, the resulting matrix element can be shown to be identical to the one given by the present model. The numerical prediction of the $K_{\ell 3}$ decay rate in the model is also quite close to the experimental value.

Encouraged by the successful features of the model for ordinary meson decays, we generalize now to the case of charmed meson decays. A charmed meson is presumed to have a structure of the type $q\bar{q}$, where one of the quarks is the charmed quark c, which we will take to be the standard GIM type[4]. The other quark could be one of the standard Gell-Mann-Zweig quarks, u,d,s. For a given angular momentum and parity, there are three charmed mesons: $D^+=c\bar{d}$, $D^o=c\bar{u}$ which form an isodoublet, and $F^+=c\bar{s}$, an isosinglet. The weak hadronic current now transforms as a 15-plet under SU(4). The effective weak Hamiltonian is given by Eq.(1), with the hadronic current now in the form

$$J_\mu = J_\mu^C + J_\mu^{GIM}$$
$$J_\mu^{GIM} = \cos\theta_c(V_{\mu 4}^3+A_{\mu 4}^3)-\sin\theta_c(V_{\mu 4}^2+A_{\mu 4}^2) \ , \tag{9}$$

where the indices 1 to 4 represent the quarks u,d,s and c, respectively, and J_μ^C is the Cabibbo current given by Eq. (2). We have not displayed color degree of freedom but shall assume that the hadronic currents transform as a singlet in color space. With the current given by Eq. (9), the non-leptonic weak Hamiltonian transforms under SU(4) as a symmetric representation in $\underset{\sim}{15} \otimes \underset{\sim}{15}$, and leaving out the singlet, as $\underset{\sim}{20}_s \oplus \underset{\sim}{84}$. Now the $\underset{\sim}{84}$-plet of SU(4) contains, besides the octet, the $\underset{\sim}{27}$-plet of SU(3) which is charm preserving. On the other hand in terms of SU(3) content $\underset{\sim}{20}_s \supset \underset{\sim}{8} + \underset{\sim}{6} + \underset{\sim}{6}^*$, where only the octet induces the strangeness changing but charm preserving non-leptonic decays. Clearly, the SU(4) representation $\underset{\sim}{20}_s$ is the analogue of the octet in SU(3) and may be expected to be dynamically enhanced compared with $\underset{\sim}{84}$, just as in SU(3) the octet is enhanced compared with the $\underset{\sim}{27}$-plet. Assuming $\underset{\sim}{20}_s$-plet dominance[11], the non-leptonic Hamiltonian may be expressed as

$$H_W^{N.L.} \sim x \frac{G_F}{\sqrt{2}} (J_\mu J_\mu^+)_{20_s} + \text{h.c.} \ , \tag{10}$$

where in an SU(4) invariant theory the enhancement factor x in Eq. (10) would be the same as the octet enhancement factor in Eq. (3). We will assume this to be the case, so that non-leptonic charmed meson decays can be computed in terms of the known $K_1^o \to 2\pi$ decay. Alternatively, the ratio of charm particle decays into different modes can be computed independently of this assumption. The charm changing piece of $H_W^{N.L.}$ in Eq. (10) transforms as a sextet under SU(3) with the Cabibbo angle favored piece ($\propto\cos^2\theta_c$) having a quark structure $(\bar{d}u)(\bar{c}s)$ + h.c. For non-leptonic charmed meson decays, we will confine our attention only to the Cabibbo angle favored piece, which imposes the selection rule

$$\Delta C = \Delta S = \pm 1 \quad , \quad |\Delta I| = 1 \ . \tag{11}$$

We now generalize the phenomenological model, taking as in Eq. (4), the vector and axial vector currents to be proportional to the vector-meson and the derivative of the pseudoscalar meson fields respectively. The coupling constants f_{D*} and f_{F*} for the charmed vector mesons D* and F* will be obtained from the vector dominated form of the first spectral function sum rule for SU(4). This leads to identical values of m_V^2/f_V^2 for ρ,K*,D* and F*. For the pseudoscalar coupling constants we will use the SU(4) symmetry result $f_D=f_F=f_K=f_\pi$. In our model, the charm changing non-leptonic interaction, like its strangeness changing counterpart, is effectively a two-meson vertex. The selection rule (11) shows that whereas the charmed particles D^o and F^+ (pseudoscalar or vector) can interact at the lowest order and Cabibbo favored weak non-leptonic vertex, the meson D^+ cannot do so. This is because positively charged K or K*'s do not occur with strangeness -1.

To describe the two-body hadronic decays of the charmed mesons in the present model, where the particles in the final state may be vector or pseudoscalar mesons, we also need the knowledge of the strong vertices containing three vector or pseudoscalar mesons. Since the three pseudoscalar meson strong vertex does not exist, at least one of the particles involved at such vertices must be a vector meson. Generalizing the concept of universality of vector couplings to SU(4), we adopt for this purpose the strong Hamiltonian

$$\begin{aligned} H_{str} &= ig\ \mathrm{Tr}(\phi_\mu P \overleftrightarrow{\partial}_\mu P) + \frac{2g}{m}\ \varepsilon_{\mu\nu\lambda\rho}\ \mathrm{Tr}(P\partial_\mu \phi_\nu \partial_\lambda \phi_\rho) \\ &\quad - \frac{2}{3}\ ig\ \mathrm{Tr}(F_{\mu\nu}\phi_\mu\phi_\nu) - \frac{2ig}{9m^2}\ \mathrm{Tr}(F_{\mu\nu}F_{\nu\lambda}F_{\lambda\mu}) \end{aligned} \tag{12}$$

where we have written down the couplings of a vector meson only to other vector or pseudoscalar mesons. In Eq. (12), g is the universal coupling constant and m is the invariant mass. The form (12) is an obvious generalization of the Sakita-Wali interaction[12] Hamiltonian relevant to SU(3). If SU(3) is any guide, we may hope that H_{str} given by Eq. (12) is not unreasonable. Experience based on SU(3) indicates that coupling constants, unlike masses, are better described by group symmetry. Besides, the vector meson couplings may be expected to have a somewhat special status embodied through universality. At any rate, considering our present state of ignorance about charmed particles, it is neither feasible nor perhaps desirable to complicate the analysis by introducing symmetry breaking effects through unknown parameters. Given H_{str} through Eq. (12), we can now compute the two-body decays in the model with final states of the type PP,VP and VV. The numerical value of the coupling g is not needed if we normalize the charmed meson decays to $K_1^o \to 2\pi$. It should be noted that the quasi-two-body decays into VP final states would manifest themselves as three or more pseudoscalar meson final states, and similarly those into VV states as four or more pseudoscalar mesons.

What about direct three or more particle decays, which do not go through quasi-two-body modes? From the average charge multiplicity observed at SLAC at energies above the presumed charm threshold, it is enough to consider only the direct three- or four-body final hadronic states. Now, there is no reliable way to calculate multihadronic decay rates, so we will resort to estimating these by current algebra techniques based on soft mesons. Gaillard et al[1] have shown that whereas the direct decay to three pseudoscalar mesons may be a sizable fraction of the two-body decays, the direct four-body decay modes are quite negligible. Confining ourselves only to three-body decays then, it should be observed that energetics would forbid all three particles from being vector mesons if the charmed particle masses are $\sim$2 GeV. For the remaining direct three-body decays, we will treat the pseudoscalar mesons as soft to estimate the three-body decay rate in terms of appropriate two-body or quasi-two-body rates, using the standard current algebra technique. The leptonic and semi-leptonic decays of charmed mesons can also be discussed in the model in a manner described before for ordinary mesons. For the semi-leptonic decays, we consider only the $P\ell^+\nu_\ell$ and $V\ell^+\nu_\ell$ modes of decay.

An important question to raise at this stage is whether the lightest charmed particles are pseudoscalar mesons, as is the case for ordinary hadrons. Now, the SU(4) mass formulas predict[1,13] that charmed pseudoscalar and vector mesons lie quite close in mass, with the pseudoscalar mesons slightly lower. However, the SU(4) mass formulas cannot be relied upon in the prediction of level ordering for near degenerate levels. We shall accordingly treat this as an open question, and discuss both the cases when the pseudoscalar or vector mesons are the lowest lying. Only the lowest lying charmed mesons would be quasi-stable decaying by weak interactions if we believe that charm is conserved in strong and electromagnetic interactions. The higher lying levels would of course decay into the lower lying ones either by strong interactions, if kinematics allows, or radiatively.

III. DECAY RATES

A partial listing of our results is displayed in Tables I and II for the decay of charmed pseudoscalar and vector mesons respectively. We have tabulated the two most prominent non-leptonic decay modes out of all two- or quasi-two-body channels, and have also listed the dominant leptonic or semi-leptonic decay mode. Current algebra estimates show that the total direct three-body decay rate is a small fraction of the total two-body and quasi-two-body rate; these results have not been included in the tables. The tables also show the partial rates for each of the prominent decay modes listed, and the sum of all calculated two-body, quasi-two-body rates as well as all the leptonic and semi-leptonic decay rates. In all calculations, the invariant mass m is taken to be the mass of the decaying particle[14]. Finally, the last columns in the tables list the 'calculated' branching ratio.

There are several striking features of our results, which we will discuss now.

1. Note in particular that our calculations show a pronounced $D^+ \to \bar{K}^{o*}\pi^+$ decay mode for the pseudoscalar meson D^+, which should produce a clear $K^-\pi^+\pi^+$ signal for experimental detection. However, Boyarski et al[15] have set a rather stringent upper limit ($\sim$7.2%) for the branching ratio of D^+ into this mode. Even with the theoretical and experimental uncertainties, the discrepancy appears to be too large. The calculated non-leptonic decays of charmed vector mesons on the other hand are consistent with the bounds given by Boyarski et al. This suggests that the pseudoscalar charmed mesons, especially D^+, cannot all be the lowest lying charmed states.

2. The calculated leptonic and semi-leptonic branching ratios of pseudoscalar charmed mesons are too small to adequately account for the rather copious production of dimuons observed in neutrino induced reactions, particularly since the presumed charm production is suppressed by the Cabibbo factor $\sin^2\theta_c$ and by the relative rarity of sea quarks. Theoretical analysis[16] shows that for the GIM model to be viable in providing an explanation of the dimuon data, the leptonic and semi-leptonic branching ratio of the charmed particle decay should be at least about 10%. Note that the helicity argument which suppresses the pure leptonic decay rate of pseudoscalar mesons,does not suppress the leptonic decays of vector mesons. In particular the leptonic decay of F^{+*} is Cabibbo angle favored and has a large branching ratio which could, in principle, explain the rather copious production of

TABLE I. Prominent charmed pseudoscalar meson decays. The The total 'calculated' decay rates are $\Gamma(D^o)=1.09\times10^{13}sec^{-1}$, $\Gamma(D^+)=3.14\times10^{12}$ sec^{-1}, $\Gamma(F^+)=1.36\times10^{13}$ sec^{-1}.

Process	$\Gamma(sec^{-1}) \times 10^{-12}$	Branching Ratio
$D^o \rightarrow \rho^+K^-$	4.72	0.43
$\rightarrow \bar{K}^{o*}\eta$	1.63	0.15
$\rightarrow K^-\ell^+\nu_\ell$	0.03	-
$D^+ \rightarrow \bar{K}^{o*}\pi^+$	1.86	0.59
$\rightarrow \rho^+\bar{K}^o$	1.15	0.37
$\rightarrow \bar{K}^o\ell^+\nu_\ell$	0.03	-
$F^+ \rightarrow \bar{K}^{o*}K^+$	3.94	0.29
$\rightarrow \rho^+\pi^o$ or $\pi^+\rho^o$	3.07	0.23
$\rightarrow \eta\ell^+\nu_\ell$	0.02	-

TABLE II. Prominent charmed vector meson decays. The total 'calculated' decay rates are $\Gamma(D^{o*})=5.01\times10^{12}$ sec^{-1}, $\Gamma(D^{+*})=3.34\times10^{12}$ sec^{-1}, $\Gamma(F^{+*})=8.32\times10^{12}$ sec^{-1}.

Process	$\Gamma(sec^{-1}) \times 10^{-12}$	Branching Ratio
$D^{o*} \rightarrow \bar{K}^{o*}\rho^o$	1.19	0.24
$\rightarrow K^-\rho^+$	0.85	0.17
$\rightarrow K^{-*}\ell^+\nu_\ell$	0.006	-
$D^{+*} \rightarrow \bar{K}^{o*}\rho^+$	1.86	0.56
$\rightarrow \pi^+\bar{K}^{o*}$	0.67	0.20
$\rightarrow \ell^+\nu_\ell$	0.09	0.03
$F^{+*} \rightarrow \rho^+\rho^o$	2.04	0.25
$\rightarrow K^+\bar{K}^{o*}$	0.71	0.09
$\rightarrow \ell^+\nu_\ell$	1.58	0.19

dimuons. It should be remarked that there may be other difficulties such as with the x and y distributions or the somewhat copious production of kaons seen in dilepton production[17] which might force one to abandon the simple GIM model, and therefore the present calculations which are based on it. However, at present, we do not see the need for this.

3. In the $e\bar{e}$ experiments, if we assume that roughly half of the post charm-threshold physics is due to production of charmed vector mesons, all three species being produced with equal probability, we find as we cross the charm threshold that the charged particle multiplicity should rise from the experimental value of about 3.5 to about 3.9, and the fraction of events with a K^- should rise from about[18] 0.2 to about 0.26. In calculating these numbers we have not folded in the effect of direct three or more-body decays and have further assumed that the charmed pair production is not accompanied by other ordinary particles like pions; both these effects would tend to raise the average charged particle multiplicity. However if the new physics involves, besides charmed particles, the production of a pair of heavy leptons[19], one would expect that at energies above the various thresholds, the calculated values of the two numbers would be somewhat smaller than the ones quoted above. If 35-40% of the new events are due to heavy leptons, the post-threshold charged particle multiplicity and the fraction of events with a K^- reduce[20] to 3.6 and 0.21 respectively. Note in particular that the fraction of events with a K^- does not change very much as the charm and heavy lepton thresholds are crossed, in conformity with the experimental data. The charged particle multiplicity is somewhat lower than what the data suggests, but the multihadronic decays neglected in this calculation, should bring this up. Finally, we would like to mention that the numbers calculated here do not change much if instead of vector mesons, the pseudoscalar charmed mesons are the lower lying ones.

IV. $K^-\pi^+$ AND $K^-\pi^+\pi^+\pi^+$ DECAY MODES

As mentioned in the introduction, recently SLAC has observed[2] narrow peaks in the mass spectra of the hadronic states $K^-\pi^+$ and $K^-\pi^+\pi^-\pi^+$, both at about 1.86 GeV. A very likely interpretation is that these hadronic states in the peak are arising from the decay of the same state with a mass $\sim$1.86 GeV, which is presumably a charmed particle state. SLAC has made a preliminary determination of the relative abundance γ of the

$K3\pi$ state over the $K\pi$ state, and obtained

$$\gamma = 3.35 {+1.85 \atop -1.11} . \qquad (13)$$

It is of special interest to see what the theoretical prediction for γ is on the basis of the present model. Since the spin-parity of the new state has not been determined yet, we shall calculate γ from both the pseudoscalar and vector meson decays. A glance at Table II shows that the dominant decay mode of the vector meson D^{o*} is to the final state $\bar{K}^{o*}\rho^o$ which would lead to the observable state $K^-\pi^+\pi^-\pi^+$, so if the new state seen at SLAC is the vector charmed meson, one would expect a lot more $K3\pi$ than $K\pi$.

For decay into the two-body states, the model gives

$$\frac{\Gamma(D^o \to \bar{K}^{o*}\rho^o)}{\Gamma(D^o \to K^-\pi^+)} = 7.1 \times 10^{-3} \qquad (14)$$

$$\frac{\Gamma(D^{o*} \to \bar{K}^{o*}\rho^o)}{\Gamma(D^{o*} \to K^-\pi^+)} = 4.2 . \qquad (15)$$

Thus if the decaying particle is a pseudoscalar D^o, the decay mode $D^o \to \bar{K}^{o*}\rho^o$ is negligibly small compared to $D^o \to K^-\pi^+$, and cannot by itself explain the rather copious $K3\pi$ state seen in the SLAC experiment. On the other hand, using the branching ratios for $\bar{K}^{o*} \to K^-\pi^+$ and $\rho^o \to \pi^-\pi^+$, we see that if the decaying particle is a D^{o*}, the production of $K^-\pi^+\pi^-\pi^+$ through the state $\bar{K}^{o*}\rho^o$ would be more copious than that of $K^-\pi^+$ by a factor of about 2.8. Now, besides $\bar{K}^{o*}\rho^o$, the hadronic state $K^-\pi^+\pi^-\pi^+$ can also arise from other direct three- or four-body states like $\rho^o K^-\pi^+$, $\bar{K}^{o*}\pi^+\pi^-$ and $K^-\pi^+\pi^-\pi^+$. For D^{o*} decay, current algebra estimates show that the contribution to $D^{o*} \to K^-\pi^+\pi^-\pi^+$ from the direct three- and four-body modes is about 35% of that from the $\bar{K}^{o*}\rho^o$ mode, so that

$$\gamma_{D*} \equiv \frac{\Gamma(D^{o*} \to K^-\pi^+\pi^-\pi^+)}{\Gamma(D^{o*} \to K^-\pi^+)} \simeq 3.8 . \qquad (16)$$

For the D^o decay, on the other hand, current algebra gives

$$\gamma_D \equiv \frac{\Gamma(D^o \to K^-\pi^+\pi^-\pi^+)}{\Gamma(D^o \to K^-\pi^+)} \simeq 0.66, \qquad (17)$$

where the dominant contribution ($\simeq$70%) comes from the decay $D^o \to \rho^o K^-\pi^+$. The experimental result (13) seems to support the alternative that SLAC is observing the decay of the vector D^{o*} rather than the pseudoscalar D^o, although it is perhaps too early to draw any definitive conclusion yet.

However, if the decaying particle is a D^{o*}, we note that the final state $K^-\pi^+\pi^-\pi^+$ should arise dominantly from the decay mode $D^{o*} \to \bar{K}^{o*}\rho^o$, so that K^- and one of the π^+ should possess an invariant mass close to the mass of the $\bar{K}^*$, while the other π^+ and the π^- should peak near the invariant mass of the ρ. On the other hand, if the weakly decaying particle is a D^o, there should be hardly any events containing both $\bar{K}^*$ and ρ, but the bulk of the contribution should come from the decay mode $D^o \to \rho^o K^-\pi^+$, so one should see the ρ with little or no $\bar{K}^*$. Evidently, experimental clarification on the resonant structure of the hadronic state $K^-\pi^+\pi^-\pi^+$ would be of great interest.

V. DISCUSSION

We have seen that the present data from several experiments seem to favor the possibility that the lowest lying charmed particles are vector mesons. This is contrary to the situation for ordinary hadrons, and if true, would imply a rather special kind of $q\bar{q}$ interaction.

Our entire discussion in the work has been based first and foremost on the validity of the simple GIM model of charm. It is conceivable that the simple model might need modification, specially if the recently conjectured[21] right-handed hadronic currents exist. Next, the phenomenological model we have used to discuss charmed meson decays, may be suspect. However, vector dominance which is an essential ingredient in this model has been with us for a long time, and has been reasonably successful in the description of ordinary hadrons. There has been some recent difficulty in accounting[22] for the observed radiative decays of vector mesons, but it is too early to tell. Without the kind of model discussed here, we do not know of any other method which could allow detailed calculations of the type we have made. In the emerging field of charmed particle decays, we feel it is better to have specific predictions in hand even though the quantitative estimates may not always be completely accurate.

It may be useful to spell out a few words of caution. Whenever a decay mode has a very small branching ratio, it is usually (not always) because of partial cancellation between various contributions to the decay amplitude, with no apparent overall simple explanation based on symmetry arguments. The cancellation, however, is sensitive to the numerical values of the parameters like f_V, f_P, m etc., so that changes in these numerical values could get magnified in such cases. In particular some ambiguity exists as to what value one should choose for the mass m appearing at some of the strong vertices through Eq. (12). We have throughout chosen m to be the mass of the decaying particle. To get an idea of how sensitive the results are to the choice of m, we have recalculated the results (14) and (15) using for m^2 the average of the squared masses of the mesons participating at the strong vertex. This changes the result (15) to 6.2 but increases (14) to 8.8×10^{-2}, which is an order of magnitude bigger but is still small compared with the result (17). The current algebra result would also change, but the quantitative reliability of the current algebra estimates may be questionable on more serious grounds that the pseudoscalar mesons treated as soft, may sometimes not be soft at all. In such cases, one can only hope that the estimates are not too unreliable. We would also like to point out that although we have throughout taken the hadronic weak currents to have a V-A structure, in those decays where there are no parity violting effects, it clearly does not matter. In particular, the result (15) is valid also for V+A hadronic currents, and besides, to a good approximation ($m_\pi^2 \to 0$), it is independent[5] of the coupling constants f_v and f_p.

REFERENCES

1. M. K. Gaillard, B. W. Lee and J. L. Rosner, Rev. Mod. Phys. 47, 277 (1975).
2. G. Goldhaber et al., Phys. Rev. Lett. 37, 255 (1976).
3. A. Benvenuti et al., Phys. Rev. Lett. 34, 419 (1975).
4. S. L. Glashow, J. Iliopoulos and L. Maiani, Phys. Rev. D2, 1285 (1970), hereafter referred to as GIM.
5. S. R. Borchardt and V. S. Mathur, Phys. Rev. Lett. 36, 1287 (1976), and Rochester preprint (1976); S. R. Borchardt, T. Goto and V. S. Mathur, to be published.
6. J. J. Sakurai, Phys. Rev. 156, 1508 (1967); G.S.Guralnik, V. S. Mathur and L. K. Pandit, Phys. Rev. 168, 1866 (1968).
7. V. Chaloupka et al., Phys. Lett. 50B, 1 (1974).
8. S. Weinberg, Phys. Rev. Lett. 18, 507 (1967); T. Das, V.S. Mathur and S. Okubo, Phys. Rev. Lett. 18, 761 (1967).
9. Y. Hara, Y. Nambu and J. Schechter, Phys. Rev. Lett. 16, 380 (1966). Note that the baryon pole terms make a vanishing contribution (in SU(3) limit) to the parity violating hyperon decays.

10. M. K. Gaillard and B. W. Lee, Phys. Rev. Lett. 33, 108 (1974); G. Altarelli and L. Maiani, Phys. Lett. 52, 351 (1974).

11. R. Kingsley, S. B. Treiman, F. Wilczek and A. Zee, Phys. Rev. D11, 1919 (1975), and 12, 106 (1975); G. Altarelli, N. Cabibbo and L. Maiani, Phys. Rev. Lett. 35, 635 (1975), and Nucl. Phys. B88, 285 (1975); Y. Iwasaki,Phys. Rev. Lett. 34, 1407 (1975); M. Einhorn and C. Quigg, Phys. Rev. D12, 2015 (1975), and Phys. Rev. Lett. 35 ,1407 (1975).

12. B. Sakita and K. C. Wali, Phys. Rev. Lett. 14, 404 (1965).

13. S. Okubo, V. S. Mathur and S. R. Borchardt, Phys. Rev. Lett. 34, 236 (1975).

14. The mass of the D or D* meson is taken to be 1.86 GeV, and the masses of F and F* are then computed from the broken SU(4) mass formula.

15. A. M. Boyarski et al., Phys. Rev. Lett. 35, 195 (1975).

16. V. Barger, R. J. N. Phillips and T. Weiler, Phys. Rev. D9, 2511 (1976).

17. J. von Krogh et al., Phys. Rev. Lett. 36, 710 (1976).

18. R. Schwitters, in Proceedings of the International Symposium on Lepton and Photon Interactions at High Energies, Stanford, California, 1975, edited by W. T. Kirk (Stanford Linear Accelerator Center, Stanford, Calif. 1975).

19. M. L. Perl, in Proceedings of the Canadian Institute of Particle Physics Summer School, Montreal, Quebec, Canada, 16-21 June 1975 (to be published); M. L. Perl et al., Phys. Rev. Lett. 35, 1489 (1975).

20. These results are based on Harari's estimates in the Proceedings of the International Symposium on Lepton and Photon Interactions at High Energies, Stanford, California, 1975, edited by W. T. Kirk (Stanford Linear Accelerator Center, Stanford, Calif. 1975).

21. See for example, R. N. Mohapatra, Phys. Rev. D6, 2023 (1972); A. DeRujula, H. Georgi and S. L. Glashow, Phys. Rev. Lett. 35, 69 (1975).

22. See for example, P. J. O'Donnell, Phys. Rev. Lett. 36, 177 (1976).

Angular Momentum in Non-Abelian Gauge Field Theories

KERSON HUANG AND DANIEL R. STUMP[+]
Laboratory for Nuclear Science and Department of Physics
Massachusetts Institute of Technology
Cambridge, Massachusetts

ABSTRACT

In Yang-Mills theories, there are contributions to the angular momentum due to the longitudinal gauge electric field produced by the sources. In the quantized theory, this can result in states with anomalous angular momentum. We present an elementary and general discussion of this point.

Recently, Jackiw and Rebbi[1], and Hasenfratz and 't Hooft[2], found spin 1/2 solutions in a Yang-Mills theory containing isospin 1/2 boson fields but no fermion fields. Apparently, in such a theory a spatial rotation can induce an isospin rotation, and thus angular momentum becomes mixed with isospin. The purpose of the present investigation is to derive an expression for the angular momentum operator that displays, in elementary and general form, the possibility of mixing internal and rotational symmetries.

We are motivated by what is well-known in electromagnetism. The angular momentum of an electromagnetic system is the sum of three parts; (i) that of the transverse radiation field, (ii) that due to motion of the sources, and (iii) that of the longitudinal field produced by the sources. The importance of part (iii) is illustrated by the Einstein-de Haas effect[3] and the charge-monopole system. Our aim is to isolate the corresponding quantity in a general gauge field theory.

Consider a standard Yang-Mills theory[4] with Lagrangian density

$$L = -\frac{1}{4} F_a^{\mu\nu} F_{a\mu\nu} + (D_\mu \phi)^* (D^\mu \phi) - V(\phi), \tag{1}$$

[+]Present Address: Department of Physics, University of Rochester, Rochester, New York where the source field ϕ is a multicomponent scalar field, and

$$F^{\mu\nu}_a = \partial^\mu A^\nu_a - \partial^\nu A^\mu_a + e\, C_{abc}\, A^\mu_b A^\nu_c ,$$
$$D^\mu = \partial^\mu - ie\, A^\mu_a L_a . \qquad (2)$$

The theory is invariant under gauge transformations

$$\phi(x) \to U(x)\phi(x),$$
$$A^\mu_a(x)L_a \to U(x)[A^\mu_a(x)L_a + \frac{i}{e}\,\partial^\mu]U^{-1}(x),$$
$$U(x) \equiv \exp[-iw_a(x)L_a] , \qquad (3)$$

where $w_a(x)$ is a real continuous function. The L_a's $(a=1,\ldots,n)$ are $m \times m$ matrices on the space spanned by the m components of ϕ, and form a representation (generally reducible) of the Lie algebra of a semi-simple group:

$$[L_a, L_b] = i\, C_{abc} L_c . \qquad (4)$$

The term $V(\phi)$ is a gauge-invariant function of ϕ, independent of derivatives of ϕ. The field equations which follow from (1) are

$$D_\mu D^\mu \phi = - \partial V/\partial\phi^*,$$
$$\partial_\nu F^{\mu\nu}_a = j^\mu_a ,$$
$$j^\mu_a \equiv [ie\, \phi^*\, L_a D^\mu \phi + \text{c.c.}] + e\, C_{abc}\, F^{\mu\lambda}_b A_{c\lambda} . \qquad (5)$$

We introduce also the gauge electric and magnetic fields $\vec{E}_a$, $\vec{B}_a$ by

$$E^k_a = - F^{ok}_a ,$$
$$B^k_a = - \frac{1}{2}\, \varepsilon^{kij}\, F^{ij}_a , \qquad (6)$$

in terms of which the second equation in (5) reads

$$\nabla\cdot\vec{E}_a = - j^o_a ,$$
$$\nabla\times\vec{B}_a = - \vec{j}_a + \frac{\partial \vec{E}_a}{\partial t} . \qquad (7)$$

It is to be noted that the first equation is an equation of constraint: the longitudinal part of $\vec{E}_a$ is not a dynamically independent field.

To determine the angular momentum, we first construct the energy-momentum tensor, $T^{\mu\nu}$. Let

$$T_o^{\mu\nu} = - g^{\mu\nu} L + \sum_F \frac{\partial L}{\partial(\partial_\mu F)} \partial^\nu F, \tag{8}$$

where F stands for all fields in L; by (1),

$$T_o^{\mu\nu} = -g^{\mu\nu}L+[(D^\mu\phi)(\partial^\nu\phi^*)+(D^\mu\phi)^*(\partial^\nu\phi)]-F_a^{\mu\rho}\partial^\nu A_{a\rho}. \tag{9}$$

The equations of motion imply that $T_o^{\mu\nu}$ is conserved; however, it is neither symmetric nor gauge invariant. The energy-momentum tensor, $T^{\mu\nu}$, is constructed according to the Belinfante method[5] by adding to $T_o^{\mu\nu}$ a tensor $T'^{\mu\nu}$ with the properties

(a) $T^{\mu\nu} = T_o^{\mu\nu} + T'^{\mu\nu}$ is symmetric,

(b) $\partial_\mu T^{\mu\nu} = 0$,

(c) $\int d^3r\, T'^{oo} = 0$.

The third property ensures that the Hamiltonian H is given by

$$H = \int d^3r\, T_o^{oo} = \int d^3r\, T^{oo} . \tag{10}$$

For the Lagrangian (1), $T'^{\mu\nu}$ is given by

$$T'^{\mu\nu} = \partial_\rho (A_a^\nu F_a^{\mu\rho}) , \tag{11}$$

which leads to the gauge invariant energy-momentum tensor

$$T^{\mu\nu} = -g^{\mu\nu}L + [(D^\mu\phi)(D^\nu\phi)^* + \text{c.c.}] - F_a^{\mu\rho}F_{a\rho}^{\nu} , \tag{12}$$

using the equation of **motion** (5).

Next, we define the angular momentum tensor

$$M^{\lambda\mu\nu} = x^\mu T^{\lambda\nu} - x^\nu T^{\lambda\mu} , \tag{13}$$

which is conserved because $T^{\mu\nu}$ is symmetric:

$$\partial_\lambda M^{\lambda\mu\nu} = T^{\mu\nu} - T^{\nu\mu} = 0. \tag{14}$$

Finally, the angular momentum J^k (latin indices run from 1 to 3) is defined by

$$J^k = \frac{1}{2}\,\varepsilon^{k\ell m}\int d^3r\, M^{o\ell m}, \tag{15}$$

i.e.

$$\vec{J} = \int d^3r\, \vec{r} \times \vec{G}\,,$$

where $G^k = T^{0k}$ is the momentum density.

$\vec{G}$ is calculated from Eq. (12) with the result

$$\vec{G} = \vec{E}_a \times \vec{B}_a - [\pi\vec{D}\phi + c.c.], \tag{16}$$

where

$$\pi = \frac{\partial L}{\partial(\partial_o\phi)} = (D^o\phi)^*,$$

$$\vec{D} = \nabla + ie\,\vec{A}_a\,L_a. \tag{17}$$

The terms in (16) are separately gauge-invariant. We now wish to separate the dynamical fields from the constrained fields. Therefore we split $\vec{E}_a$ into transverse and longitudinal parts

$$\vec{E}_a = \vec{E}_a^{\,T} + \vec{E}_a^{\,L},$$

$$\nabla\cdot\vec{E}_a^{\,T} = 0,$$

$$\vec{E}_a^{\,L} = -\nabla\Omega_a, \tag{18}$$

where Ω_a is determined by (7). With these definitions, and again using the field equation (7), the momentum density becomes

$$\vec{G} = \vec{E}_a^{\,T}\times(\nabla\times\vec{A}_a) - [\pi\nabla\phi + c.c.]$$

$$+\vec{E}_a^{\,L}\times(\nabla\times\vec{A}_a) - \vec{A}_a(\nabla\cdot\vec{E}_a^{\,L}). \tag{19}$$

The angular momentum density is $\vec{M}\equiv\vec{r}\times\vec{G}$; straightforward calculation shows that

$$\vec{M} = \vec{M}_o + \vec{M}_1, \tag{20}$$

where

$$\vec{M}_o = r\times[\vec{E}_a^{\,T}\times(\nabla\times\vec{A}_a) - (\pi\nabla\phi + c.c.) + \Omega_a\nabla(\nabla\cdot\vec{A}_a)],$$

$$\vec{M}_1 = -\partial_i[E_a^{Li}(\vec{r}\times\vec{A}_a)] - \partial_i[\Omega_a(\vec{r}\times\nabla)A_a^i] - \nabla\times(\Omega_a\vec{A}_a). \tag{21}$$

The angular momentum is

$$\vec{J} = \vec{J}_o + \vec{J}_1,\quad \vec{J}_o = \int d^3r\,\vec{M}_o,\quad \vec{J}_1 = \int d^3r\,\vec{M}_1. \tag{22}$$

So far, no choice of gauge has been made. Now we go to the Coulomb gauge: $\nabla\cdot\vec{A}_a=0$ $(a=1,\ldots,n)$. Then $\vec{J}_o$ involves only dynamical fields, and we recognize it as the angular momentum for a non-interacting field theory (i.e. for e=0) involving the fields which are present in L. For that reason, $\vec{J}_o$ has only integer eigenvalues. On the other hand, $\vec{J}_1$ depends on the longitudinal field $\vec{E}_a{}^L$. The existence of 1/2 integer eigenvalues of $\vec{J}$, and mixing of angular momentum with isospin, can only be due to the term $\vec{J}_1$.

It remains to simplify the formula for $\vec{J}_1$. By Gauss's theorem, $\vec{J}_1$ can be written as a pure surface integral

$$\vec{J}_1 = -\int dS\{\hat{n}\cdot\vec{E}_a{}^L(\vec{r}\times\vec{A}_a)+(\hat{n}\times\vec{A}_a)\Omega_a+\hat{n}^j\Omega_a(\vec{r}\times\nabla)A_a^j\}, \tag{23}$$

where dS is an element of a closed surface at infinity, $\hat{n}$ the outward normal. Note that $\vec{J}_1$ is zero unless there are long-ranged fields present. To proceed further, consider a particular component of $\vec{J}_1$, say J_1^3 , and choose as integration surface an infinitely long cylinder with axis along the z-axis, and radius $R\to\infty$. With this choice, J_1^3 becomes

$$J_1^3 = -\int_{-\infty}^{\infty} dzR\int_o^{2\pi} d\theta\,[\hat{n}\cdot\vec{E}_a^L(\vec{r}\times\vec{A}_a)^3+\Omega_a(\vec{r}\times\nabla)^3\hat{n}\cdot\vec{A}_a]. \tag{24}$$

On this cylinder surface, the Coulomb potential Ω_a is

$$\Omega_a = \frac{1}{4\pi}\int dz'd^2x_\perp'(-j_a^o(z',\vec{x}_\perp'))[(z-z')^2+(\vec{x}_\perp-\vec{x}_\perp')^2]^{-\frac{1}{2}}$$

$$\xrightarrow[R\to\infty]{} -\frac{1}{4\pi}\int dz'\ \sigma_a(z')[(z-z')^2+R^2]^{-\frac{1}{2}}, \tag{25}$$

where

$$\sigma_a(z) = \int d^2x_\perp\ j_a^o(z,\vec{x}_\perp)\ . \tag{26}$$

The important property of (25) is that Ω_a, and consequently $\vec{E}_a{}^L$, are independent of the azimuthal angle on the surface of the cylinder. Therefore the second term in (24) vanishes by the continuity of $\hat{n}\cdot\vec{A}_a$ around the cylinder, leaving only the first term which can be recast as

$$J_1^3 = \int dz\ R\,\frac{\partial\Omega_a(R,z)}{\partial R}\,\Phi_a^3(R,z), \tag{27}$$

where $\Phi_a^3(R,z)$ is the total flux of $\nabla\times\vec{A}_a$ through the cylinder at the level z:

$$\Phi_a^3(R,z) = \int_0^{2\pi} d\theta(\vec{r}\times\vec{A}_a)^3 = \oint_{C_3(R,z)} d\vec{s}\cdot\vec{A}_a$$

$$= \int_{D_3(R,z)} d^2x_\perp (\nabla\times\vec{A}_a)^3, \tag{28}$$

where $C_3(R,z)$ is a circle of radius R at the level z, parallel to the xy-plane, and $D_3(R,z)$ is the disc bounded by C_3. Finally, in the limit $R\to\infty$, $\Phi_a^3(R,z) \to \Phi_a^3$, a constant independent of z, by conservation of flux. Thus if we assume that the limit $R\to\infty$ can be taken inside the integral in (27), we obtain

$$J_1^3 = \frac{1}{2\pi}\Phi_a^3 Q_a , \tag{29}$$

where

$$Q_a = \int dz\, \sigma_a(z) = -\int d^3r\, \nabla\cdot\vec{E}_a . \tag{30}$$

Or

$$J_1^k = \frac{1}{2\pi}\Phi_a^k Q_a, \tag{31}$$

for an arbitrary component of $\vec{J}_1$. To repeat, Φ_a^k is the total flux of $\nabla\times\vec{A}_a$ in the kth direction. For non-Abelian gauge fields, this is not the same as the flux of $\vec{B}_a$. By (30), Q_a is the a charge.

Equation (31) was derived in classical field theory. We can immediately pass on to the quantum theory by replacing all fields by operators with canonical equal-time commutation relations. In sectors of the Hilbert space of the quantum theory where the direction of $\vec{\Phi}_a$ depends on internal symmetry index a, there is a mixing of angular momentum with internal quantum numbers.

We mention two examples. The first is scalar electrodynamics with spontaneous symmetry breaking. This is an Abelian theory, and all finite-energy solutions in three space dimensions give $\vec{J}_1=0$. However, Nielsen and Olesen[6] have exhibited a vortex solution with cylindrical symmetry, that may be looked upon as a finite-energy solution in two space dimensions. By general arguments[7], the flux through the vortex must obey the quantization condition $Q\Phi=2\pi n$, where n is an integer. Hence $J_1=n$. The work of Nielsen and Olesen demonstrates that there is at least one classical solution with $\Phi\neq 0$.

Next we mention the monopole solution, in an SU(2) Yang-Mills-Higgs theory, of 't Hooft[8] and Polyakov[9]:

$$A_a^i = \varepsilon^{aij} x^j \frac{A(r)}{r} \quad (a=1,2,3),$$

$$A(r) \xrightarrow[r\to\infty]{} -\frac{1}{er} . \tag{32}$$

This gives

$$\Phi_a^k = \delta_{ak} \frac{2\pi}{e} , \tag{33}$$

$$J_1^k = \frac{Q_k}{e} = I_k , \tag{34}$$

where I_a is the isospin, and the last equality follows from (30). Although (32) is a classical solution, the result (34) can be taken over directly in the quantum theory of the monopole[10]. By incorporating scalar fields of isospin 1/2 into the theory, as is done in Refs. 1 and 2, it is clear that there are states for which (34) assumes half-integer values. This leads to the possibility of constructing fermions in a field theory involving only boson fields[11].

REFERENCES

1. R. Jackiw and C. Rebbi, Phys. Rev. Lett. 36, 1116 (1976).
2. P. Hasenfratz and G.'t Hooft, Phys. Rev. Lett. 36, 1119 (1976).
3. See, e.g.,S.J.Barnett, Revs. Mod. Phys. 7, 129 (1935).
4. For a comprehensive review, see E.S. Abers and B.W. Lee, Physics Reports 9C, 1 (1973).
5. See, e.g., G. Wentzel, Quantum Theory of Fields (Interscience, New York, 1949), Appendix I.
6. H. B. Nielsen and P. Olesen, Nucl. Phys. B61, 45 (1973).
7. N. Byers and C.N.Yang, Phys. Rev. Lett. 7, 46 (1961).
8. G.'t Hooft, Nucl. Phys. B79, 276 (1974).
9. A. M. Polyakov, JETP Lett. 20, 194 (1974).
10. E. Tomboulis and G. Woo, Nucl. Phys. B107,221 (1976).
11. A. S. Goldhaber, Phys. Rev. Lett. 36, 1122 (1976).

Geometry and Interactions of Instantons

F. WILCZEK*
Joseph Henry Laboratories
Princeton University
Princeton, New Jersey

ABSTRACT

Some work on classical gauge fields is summarized. First we show how Riemannian geometry structures lead to SO(4) gauge structures, and give some applications of this idea. Second, we show how the existence of inequivalent embeddings of SU(2) in larger groups has some remarkable implications for the interactions of instantons (and monopoles).

Recently there has been a lot of interest in classical gauge fields[1-8]. In this note I will briefly summarize some of my recent work on classical gauge fields, inspired by the above papers. Two subjects are treated:

a) It is shown (what is well-known to relativists) how a Riemann structure gives rise to an SO(4) gauge structure. The instanton fits into this picture very neatly. Any solution of the Einstein equations in vacuum, even with a cosmological term, gives a solution of the SO(4) gauge field equations. We thereby get an interplay - the ideas developed for Yang-Mills theories may be applicable to gravity; the very extensive results on solutions of Einstein's equations are immediately relevant to Yang-Mills theory. So far the results of this interplay have been rather modest, but I have high hopes for it.

b) The trivial fact that SU(2) may be embedded in inequivalent ways inside a larger group has some rather non-trivial implications for the interactions of instantons. In particular, in the physically interesting case of SU(3) we find that even though <u>two</u> instantons of unit charge always repel (we are here regarding instantons as particles in a four-dimensional space) one can bring together <u>four</u> instantons, and when they sit all on top of one another, the total interaction energy is zero.

---

*Research supported in part by ERDA Contract E(11-1)3072.

Similar results hold in some models with magnetic monopoles.

I. GEOMETRY

1. The relationship between Riemannian geometry and SO(4) gauge theory is brought out by the tetrad formalism[9]. This is the formalism used to introduce spinor fields in general relativity. We summarize what little we need.

(i) Given a metric $g_{\mu\nu}(x)$ one can erect an orthonormal set of vectors $e^a_\mu(x)$ at each point so that

$$\eta_{ab} e^a_\mu(x) e^b_\nu(x) = g_{\mu\nu}(x) \tag{1}$$

$$g^{\mu\nu}(x) e^a_\mu(x) e^b_\nu(x) = \eta^{ab} , \tag{2}$$

where η_{ab} is the flat-space metric for the Euclidean spaces of most interest here (signature + + + +) η_{ab} is simply the Kronecker delta.

There is a non-uniqueness in the choice of $e^a_\mu(x)$ satisfying the above requirements. Indeed, we can change to

$$e'^a_\mu(x) = L^a_b(x) e^b_\mu(x), \tag{3}$$

as long as

$$\eta_{cd} L^a_c(x) L^b_d(x) = \eta_{ab}. \tag{4}$$

In the case at hand Eq. (4) says simply that we can make a position-dependent SO(4) rotation. In this way the Riemann metric leads to an SO(4) gauge structure. The link is made precise by:

(ii) Suppose $\psi(x)$ is a field which under the transformation (3) transforms as

$$\psi'(x) = R(L(x))\psi(x), \tag{5}$$

with R some representation of SO(4). Since the rotation L(x) may depend on position, the derivative of $\psi(x)$ does not have any simple transformation property. To remedy this, we introduce as usual a gauge field. The crucial point is that <u>the required gauge field can be determined entirely in terms of the tetrads $e^a_\mu(x)$</u>. The formula is

$$A_{\mu ab} = \{e^\nu_a(\partial_\mu e_{b\nu} - \partial_\nu e_{b\mu}) + \frac{1}{2} e^\rho_a e^\sigma_b (\partial_\sigma e_{c\rho} - \partial_\rho e_{c\sigma}) e_{c\mu}\}_{[a,b]} \tag{6}$$

where [a,b] means to antisymmetrize on a and b. One verifies directly that $A_{\mu ab}$ transforms as a gauge field should under the transformation (5).

Equation (6) is not as miraculous as it seems at first sight -it is derived by requiring that the tetrads $e^a_\mu(x)$, which are vectors both under the local gauge transformations and under general coordinate transformations, should have zero totally (gauge + Riemann) covariant derivative, i.e.

$$D_\nu e^a_\mu \equiv \partial_\nu e^a_\mu + A_{\nu ab} e^b_\mu - \Gamma^\sigma_{\nu\mu} e^a_\sigma = 0 \ . \tag{7}$$

2. From Eq. (7) it follows that there is a close relationship between the field strength in the Yang-Mills sense of our gauge field $A_{\mu ab}$ and the Riemann curvature tensor.

Indeed, if we use the relationship between the commutator of covariant derivatives and the field strength and Riemann tensors, we get by differentiating the vector v^ρ in different ways, using (7)

$$\begin{aligned} [D_\mu, D_\nu](e^a_\rho v^\rho) &= F^{ab}_{\mu\nu}(e^b_\rho v^\rho) \\ e^a_\rho [D_\mu, D_\nu] v^\rho &= e^a_\rho R_{\mu\nu}{}^\rho{}_\sigma v^\sigma \end{aligned} \tag{8}$$

or simply

$$F^{ab}_{\mu\nu} = e^a_\rho e^b_\sigma R_{\mu\nu}{}^{\rho\sigma} \tag{9}$$

Similarly the gauge field equation for no sources

$$D_\mu F^\mu{}_{\nu ab} = 0, \tag{10}$$

is equivalent to

$$D_\mu R^\mu{}_{\nu\rho\sigma} = D_\rho R_{\nu\sigma} - D_\sigma R_{\nu\rho} = 0 \tag{11}$$

where the Bianchi identity has been used.

From this equivalence it follows in particular that <u>any solution of the usual Einstein equations (possibly with a cosmological constant) gives a solution of the SO(4) gauge field equations</u>. Equation (11) has been studied for its own sake in alternative theories of gravitation[10].

3. The solutions we have found live on curved spaces - if the Riemann curvature vanishes so does the gauge field strength. If we are interested in solutions of the Yang-Mills equations on a flat space, we must go a step further. One thing we can do is exploit the conformal invariance of the Yang-Mills Lagrangian.

It is easy to see that the action

$$L = \int d^4x \sqrt{g}\, g^{\alpha\gamma} g^{\beta\delta} F_{\alpha\beta} F_{\gamma\delta}$$

is left unchanged by the conformal transformation:

$$g'_{\mu\nu}(x) = \rho^{-2}(x) g_{\mu\nu}(x); \quad A'_\mu(x) = A_\mu(x). \tag{12}$$

Thus, if we have a conformally flat space (which is not actually flat) we get a non-trivial gauge field by the construction of point 2, and if this field is source-free it will remain so when we change the metric back to that of flat space. We are therefore interested in conformally flat spaces satisfying Eq. (11). It is possible to prove the following:

Theorem. A conformally flat space satisfies Eq. (11) if and only if its scalar curvature R is constant.

It is, of course, known how the scalar curvature changes under a conformal transformation[11]. Demanding that a conformal transformation of flat space lead to a space with constant R leads to the equation

$$\partial^2 \rho = K\rho^3, \tag{13}$$

for the conformal factor ρ, where K is a constant at our disposal - i.e., to the equation of massless ϕ^4 theory!

The corresponding A_μ , according to Eq. (6) (with $e_a^\nu = \rho\delta_{a\nu}$ etc.)

$$A_{\mu ab} = \rho^{-1}(\delta_{b\mu}\partial_a\rho - \delta_{a\mu}\partial_b\rho) . \tag{14}$$

Unfortunately not much seems to be known about the solution of Eq. (13). I will not dwell too long on my long labors to find solutions of (13). It is, in fact, possible to find many solutions, but most do not seem to have much interest in the present context - they have singularities, or ρ changes sign, etc. For example, choose the conformal factor ρ to depend only on the four-dimensional radius R. Then one can solve (13) for ρ to get

$$\rho(R^2) = g(\ell n R^2) R^{-1};$$
$$g'' - g = Kg^3 \tag{15}$$

which can be solved for g in elliptic functions. To wipe out the singularity at R=0 we want g to vanish at $-\infty$, but then the periodicity of elliptic functions in general generates other zeroes.

4. There is one well-known conformally flat space of constant scalar curvature - the sphere. This corresponds to[12]

$$\rho(x^2) = \frac{a}{1 + bx^2} \quad , \tag{16}$$

$$A_{\mu ab} = \frac{b(x_a \delta_{\mu b} - x_b \delta_{\mu a})}{1 + bx^2} \quad , \tag{17}$$

$$F_{\mu\nu ab} = \frac{2b(\delta_{a\mu}\delta_{b\nu} - \delta_{a\nu}\delta_{b\mu})}{(1 + bx^2)^2} \quad . \tag{18}$$

This gauge field is the instanton[4]. So the instanton arises from the conformal (stereographic) projection of the gauge structure associated with the metric of the sphere. Of course, the metric of the sphere is highly symmetric, admitting an O(5) group of isometries. These transformations are just those found by Jackiw and Rebbi[7] by explicit calculation, and explains why the solution admits these symmetries.

5. The topological quantum numbers associated with a gravity field are

$$\int \sqrt{g}\, R_{\alpha\beta}{}^{\gamma\delta} R_{\rho\sigma}{}^{\tau\nu}\, \varepsilon^{\alpha\beta\rho\sigma}\, \varepsilon_{\gamma\delta\tau\nu} \, , \tag{19}$$

$$\int \sqrt{g}\, R_{\mu\nu\alpha\beta} R^{\mu\nu}{}_{\gamma\delta}\, \varepsilon^{\alpha\beta\gamma\delta} \, . \tag{20}$$

The first of these, the Euler number, is special to SO(4) gauge fields while the second, the Pontryagin index, exists for all gauge fields. The instanton has unit Euler index and zero Pontryagin index, as may be easily verified. If the SO(4) gauge field derived in point 3 is decomposed into its SU(2) pieces (using the isomorphism SU(2)×SU(2)=SO(4)) we get SU(2) fields with equal and opposite unit Pontryagin indices.

It would be of some interest to know what space, if any, corresponds to unit Pontryagin and zero Euler indices.

6. Although the connection drawn above between geometry and instantons has a certain esthetic appeal and may help in visualizing complicated field configurations, the more interesting implication to me is a qualitative and imprecise one. The outcome of Refs. 5-7 is that the vacuum of a gauge theory must be considered as a coherent superposition of configurations with

different Pontryagin indices, i.e. different topologies of gauge fields. Since gravity is in the sense outlined above also a gauge theory, we must expect a similar situation to exist there. This point of view has been advocated by Wheeler[13] for a long time - that the topology of space itself fluctuates and the vacuum should contain a coherent superposition of spaces of different topologies. Unfortunately, it is impossible to make these considerations precise in the absence of a reliable quantum theory of gravitation.

II. INTERACTIONS

1. For non-Abelian gauge fields based on SU(n) in four (Euclidean) dimensions the topological charge is given by[4]

$$q = \frac{1}{8\pi^2} \int F^a_{\mu\nu} \tilde{F}^a_{\mu\nu} \, d^4x \tag{21}$$

where a runs over the Lie algebra indices and

$$\tilde{F}^a_{\mu\nu} = \frac{1}{2} \varepsilon_{\mu\nu\rho\sigma} F^a_{\rho\sigma} \tag{22}$$

is the dual of F. It may be proved that (21) takes only integer values. One easily proves that the action is related to the charge by an inequality

$$\frac{1}{4} \int F^a_{\mu\nu} F^a_{\mu\nu} \geq \frac{1}{4} \left| \int F^a_{\mu\nu} \tilde{F}^a_{\mu\nu} \right| = 2\pi^2 |q| \tag{23}$$

which is saturated if and only if

$$F^a_{\mu\nu} = \pm \tilde{F}^a_{\mu\nu}. \tag{24}$$

Equation (24) was solved by the authors of Ref. 4 in the case of SU(2), q=1 who found the solution

$$A_\mu = \frac{x^2}{b^2+x^2} g^{-1}(x) \partial_\mu g(x) \tag{25}$$

where $g(x)=(x_4+i\vec{x}\cdot\vec{\sigma})/|x|$, with $\vec{\sigma}$ the Pauli matrices. By a gauge transformation we can bring A_μ into the form[7]

$$A_\mu = \frac{-b^2}{b^2+x^2} (\partial_\mu g) g^{-1} \tag{26}$$

which shows better how the charge is localized at x=0.

2. For the applications of instantons it is important to know their interactions with one another, i.e. the action of configurations with $q>1$. Especially interesting is the case of SU(3) for the description of the strong interactions.

The inequality (23) shows that the interaction energy of a gas of instantons is always positive. Great efforts have been expended by many people to find solutions with $q\geqslant 2$, with no success. It seems therefore very likely that two instantons in SU(2) or SU(3) always repel.

There is one very simple way of generating high charge solutions in larger groups than SU(2). Let us write (26) in component form

$$A_\mu = A_\mu^i \frac{\sigma_i}{2}, \tag{27}$$

leading to

$$F_{\mu\nu} = F_{\mu\nu}^i \frac{\sigma_i}{2}. \tag{28}$$

Now as long as the matrices τ^i obey the commutation relations of SU(2), the potential $A_\mu^i \tau^i$ will generate the field strength $F_{\mu\nu}^i \tau^i$, which will be self-dual. For instance, we might take instead of half the Pauli matrices, the spin-1 matrices

$$\tau_1 = \begin{pmatrix} 0 & \sqrt{2}/2 & 0 \\ \sqrt{2}/2 & 0 & \sqrt{2}/2 \\ 0 & \sqrt{2}/2 & 0 \end{pmatrix}$$

$$\tau_2 = \begin{pmatrix} 0 & -i\sqrt{2}/2 & 0 \\ i\sqrt{2}/2 & 0 & -i\sqrt{2}/2 \\ 0 & i\sqrt{2}/2 & 0 \end{pmatrix}$$

$$\tau_3 = \begin{pmatrix} 1 & 0 & 0 \\ 0 & 0 & 0 \\ 0 & 0 & -1 \end{pmatrix} \tag{29}$$

It is easy to see that this leads to <u>charge $q=\pm 4$ instantons with no interaction energy</u>!

By generalizing this construction one finds many no-interaction charges in larger groups, e.g.

$$\mathrm{SU}(3):\ \pm 1, 4 \tag{30}$$

$$\mathrm{SU}(4):\ \pm 1, 2, 4, 10 \tag{31}$$

SU(8): ±1,2,4 (two ways), 5,6,8,9,10,11,12
14,20 (two ways), 21,24,35,36,56,84. (32)

3. From this simple idea some striking conclusions follow:

a) As mentioned above, it is highly probable that two instantons of unit charge always repel (in SU(3)). However, the action is quartic in the gauge potentials so there is the possibility of three and four instanton interactions. What we have found is that, perhaps unfortunately, these complicated forces introduce qualitatively new effects.

b) The charge 4 instanton can be reached continuously from 4 separated charge 1 instantons[14]. Since at large distances, as is obvious from Eq. (26), only the two-instanton forces are important (and they are always repulsive) while when all four instantons sit on top of one another the interaction energy is zero, there must be a domain of attraction.

c) The fact that in SU(8) there are essentially different solutions for charge 4 and for charge 20 illustrates that the topological charge is insufficient to classify solutions, even after the gauge, scale, and Poincare transformations are accounted for.

4. Several interesting problems arise from these observations:

a) It would be informative and not too difficult to extend the ingenious calculations of Refs. 5, 7, 8 to the new instanton in SU(3).

b) It would be most interesting to know if the charge 4 solution in SU(3), which classically is degenerate with 4 separated unit charges, goes up or down in relative action after quantum corrections. At least the first corrections should be calculable.

c) Since viewed from the perspective of five-dimensional space-time the instanton is a static solution, just like the magnetic monopoles[1,2] in 3 space dimensions, it is natural to expect that a similar phenomenon - a spectrum of classically stable monopoles - might exist. Although the scalar Higgs fields complicate matters and the results seem to depend on the details of their potential, we have found models in which this indeed occurs.

I expect to have definite answers to a, b, and more detail on c in the very near future.

REFERENCES

1. G. 't Hooft, Nucl. Phys. B79, 276 (1974).
2. A. Polyakov, JETP Lett. 20, 194 (1974).
3. A. Polyakov, Phys. Lett. 59B, 82 (1975).
4. A. Belavin, et al., Phys. Lett. 59B, 85 (1975).
5. G. 't Hooft, Phys. Rev. Lett. 37, 8 (1976) and Harvard Preprint (1976).
6. C. Callan, Jr., R. Dashen, D. Gross, Princeton preprint (1976).
7. R. Jackiw, C. Rebbi, M.I.T. preprints (1976).
8. F. Ore, Jr., M.I.T. preprint (1976).
9. See e.g., S. Deser, P. Van Niewenhuizen, Phys. Rev. D10, 401 (1974).
10. See e.g., C. Kilmister in Perspectives in Geometry and Relativity, ed. Hoffman, Indiana University Press, Bloomington.
11. See e.g., A. Petrov, Einstein Spaces, Pergamon, N.Y., p.257-9.
12. Any book on complex variables describes the stereographic projection of the 2-sphere on the plane. The generalization to four dimensions is immediate (that's where this ρ comes from).
13. See e.g., C. Misner, K. Thorne, J. Wheeler, Gravitation, W. H. Freeman and Co., San Francisco, Chapters 43 and 44.
14. F. Wilczek, Princeton preprint (1976).

The Demise of Light Cone Field Theory (?)

C. R. HAGEN*
Department of Physics and Astronomy
University of Rochester
Rochester, New York

I. INTRODUCTION

A quick survey of papers dealing with the quark binding problem cannot help but leave one with the distinct impression that a significant number of the results obtained to date are remarkably dependent upon the techniques of light cone field theory. I am going to suggest to you here that this represents a somewhat unfortunate situation for quark binding enthusiasts because of the rather uncertain foundations which underlie light cone physics. Since this latter field has not in the relatively few years since its inception been seriously criticized, I can expect my remarks here concerning the shortcomings of light cone field theory to be considered as being little short of heretical (particularly since I see no escape from the conclusion that virtually all prior work in this area must be completely abandoned). Because of this I would like to present my observations to you in the simplest possible terms and shall hope thereby to convince you that light cone physics has indeed a most questionable future.

Not surprisingly most discussions of light cone field theories have been within the context of a four dimensional space-time. Since our world is characterized by such a dimensionality it is clear that at some point one must come to terms with the problems peculiar to four dimensions. On the other hand, it is equally true that it is virtually impossible to begin such a discussion without considerable attention being given to the more manageable two dimensional case. This is intimately related to the fact that the only manifestly covariant field theory known at the present time is the massless spinor field in two dimensions. By presenting a careful analysis of the defects of light cone field theory in two dimensions the extension to the more realistic four dimensional case can be accomplished with little or no further calculation.

---

*Supported by the U.S. Energy Research and Development Administration.

II. THE FREE SPINOR FIELD

Let us begin by discussing the case of a free spinor field in two dimensions. Light cone coordinates may be introduced for an arbitrary vector a^μ by

$$a^\pm = \frac{1}{\sqrt{2}} (a^1 \pm a^o) \tag{1}$$

so that the scalar product between two-vectors becomes

$$\begin{aligned} x^\mu y_\mu &= x^1 y^1 - x^o y^o \\ &= x^+ y^- + x^- y^+ . \end{aligned}$$

This is more simply written, of course, in terms of the metric tensor $g^{\mu\nu}$ which clearly has the form

$$g^{\mu\nu} = \begin{pmatrix} 0 & 1 \\ 1 & 0 \end{pmatrix} .$$

Since we shall adhere to a strict action principle approach we begin with the Lagrangian

$$L = \frac{i}{2} \psi\alpha^\mu \partial_\mu \psi - \frac{m}{2} \psi\beta\psi \tag{2}$$

appropriate to a free Hermitian spinor field of mass m. As is well known in two dimensions the field ψ has two components and we can consequently employ the representation

$$\alpha^+ = \sqrt{2} \begin{pmatrix} 1 & 0 \\ 0 & 0 \end{pmatrix} \quad \alpha^- = \sqrt{2} \begin{pmatrix} 0 & 0 \\ 0 & -1 \end{pmatrix} \quad \beta = \begin{pmatrix} 0 & -i \\ i & 0 \end{pmatrix}$$

for the matrices which appear in (2). These may be inferred from a Majorana representation of the Dirac matrices in the usual space-time description together with the definition (1).

To this point no attempt has been made to distinguish between the two coordinates x^+ and x^-. However, it is clear that their symmetrical status cannot be preserved indefinitely -- that we must make a statement as to whether x^+ or x^- is to serve as a time parameter. It is usual to allow x^+ to serve in that capacity, a convention which we shall adopt throughout the remainder of this talk. With this choice one then has in general a set of first order differential equations with Dirichlet boundary conditions being given on the surface x^+=constant. In conventional coordinates the corresponding equations are second order hyperbolic differential equations which are completed by the specification of Cauchy data on the surface x^o=constant. It is to be hoped that this remarkable contrast between the (classical) boundary value problems in these two different descriptions will serve in some way to make more compelling the striking deficiencies of the quantized version of the light cone theory.

Returning again to the spinor model under consideration one infers from (2) that ψ satisfies the equation

$$(\alpha^{\mu} \frac{1}{i} \partial_{\mu} + m\beta)\psi = 0$$

or in component form

$$\sqrt{2}\, \partial_{+}\psi_1 = -m\psi_2 \tag{3}$$

$$\sqrt{2}\, \partial_{-}\psi_2 = -m\psi_1 \;. \tag{4}$$

In order to simplify discussion for a moment we take m=0 and observe that the equation

$$\partial_{-}\psi_2 = 0 \tag{5}$$

allows one to infer the vanishing of ψ_2 while

$$\partial_{+}\psi_1 = 0 \tag{6}$$

remains an equation of motion for the dynamical variable ψ_1. This seemingly asymmetrical treatment of the two apparently symmetrical equations (5) and (6) is, of course, a consequence of the choice of x^+ as the "time" coordinate. It may be helpful to remark also that the vanishing of ψ_2 is analogous to the fact that in free field radiation gauge electrodynamics the equation

$$-\nabla^2 A^o = 0$$

implies that

$$A^o = 0.$$

Before proceeding further we observe that Eqs. (5) and (6) show clearly the rather general phenomenon that light cone quantization requires only half the number of dynamical variables required in conventional coordinates. (We remind those not familiar with two dimensional field theory that time derivatives appear in the equations for ψ_1 <u>and</u> ψ_2 in conventional coordinates.) The fact that ψ_2 is a dependent variable on the light cone is a peculiarity of which one must be continually aware if certain apparent contradictions are to be avoided.

An illustration of such a paradox is afforded by the consideration of the scalar

$$\sigma = \frac{1}{2}\, \psi\beta\psi.$$

Since $\psi_2=0$ for m=0 one readily infers from the nondiagonal form of the matrix β that $\sigma=0$. On the other hand the propagator for the σ field is usually obtained in terms of the fermion propagator G by computing[1]

$$G(p) = -\frac{i}{2}\int \frac{dk}{(2\pi)^2}\, \mathrm{Tr}[G(k+\tfrac{p}{2})G(k-\tfrac{p}{2})] \qquad (7)$$

a form which yields a nonvanishing result! One thus has the contradiction of the formal result $\sigma=0$ with the result (7) of a calculation. I suspect that there is at least a minority (majority?) which would view this contradiction and suggest that operator equations are notoriously singular so perhaps we shouldn't trust the result $\sigma=0$ to any great extent. This view is, however, entirely erroneous (as we shall see) and it is in fact the result (7) which must be rethought.

To display the correct version of (7) it is convenient to add the term

$$\frac{1}{2}\,\phi\psi\beta\psi$$

to the L of Eq. (2) and define matrix elements of σ as variational derivatives with respect to the external source ϕ. Since the equation for ψ_2 becomes

$$\sqrt{2}\,\partial_-\psi_2 = \phi\psi_1$$

it follows that

$$\frac{\delta\psi_2(x)}{\delta\phi(x')} = \frac{1}{2\sqrt{2}}\,\varepsilon(x^- - x^{-\prime})\psi_1(x')\delta(x^+ - x^{+\prime}).$$

Straightforward use of the action principle then yields

$$\begin{aligned} G(x-x') &= i\langle 0|(\sigma(x)\sigma(x'))_+|0\rangle = -\frac{i}{2}\mathrm{Tr}\,\frac{\delta}{\delta\phi(x')}G(x,x)\Big|_{\phi=0} \\ &\quad + \frac{i}{2\sqrt{2}}\,\varepsilon(x^- - x^{-\prime})\delta(x^+ - x^{+\prime})\langle 0|\psi_1(x)\psi_1(x')|0\rangle \\ &= -\frac{i}{2}\,\mathrm{Tr}G(x-x')G(x'-x) \\ &\quad + \frac{1}{4}\,\varepsilon(x^- - x^{-\prime})\delta(x^+ - x^{+\prime})\,\mathrm{Tr}\,G(x-x')\beta\alpha^+ \end{aligned}$$

or

$$G(p)=-\frac{i}{2}\int\frac{dk}{(2\pi)^2}\,\mathrm{Tr}[G(k+\tfrac{p}{2})G(k-\tfrac{p}{2})-\frac{1}{\left(k_- -\frac{p_-}{2}\right)}G(k+\tfrac{p}{2})\beta\alpha^+].$$

If one inserts the correct fermion propagator (which we shall subsequently derive) into this expression it is easily verified that a complete cancellation occurs. Thus results such as (5) are entirely consistent provided that one remembers as in this calculation that field theory is more than a set of Feynman rules! The latter while useful are certainly no substitute for understanding precisely what one is doing in a given calculation.

III. COVARIANCE PROBLEMS

It is customary (though by no means essential) to associate the property of covariance with the study of field theory. Thus it has happened that through long experience with conventional field theories we have good reason to believe that if we start from an apparently covariant Lagrangian the end result of any calculation will be consistent with the requirements of Poincaré invariance. In the event that we have not personally verified the formal operator covariance of a theory prior to a calculation, there is an abundance of proofs in the literature which demonstrate the covariance of virtually all of the commonly employed field theories. Thus it has become fairly commonplace to overlook entirely the question of a formal proof of covariance when one starts from an apparently covariant Lagrangian and to proceed directly to the calculation of the relevant amplitudes of the theory.

While it is potentially dangerous to implement such a view, it is probable I believe that no difficulty is likely to occur so long as one is working with conventional quantization schemes. On the other hand there is no basis for extending this hope to the realm of light cone quantization where the amount of consideration which has been given to covariance considerations is much more limited. In fact a survey of such calculations shows that each suffers from either a neglect of surface terms or a failure to consider the transformations of anything more than a subset of the total number of fields in the theory under consideration. Thus we must take up the question of whether a light cone theory is covariant with a complete lack of prejudice as to the need for conformity with "well known" results.

With these background remarks let's remind ourselves as to the precise requirements imposed by covariance. Schwinger has eloquently instructed us as to the importance of finding operators P^μ and $J^{\mu\nu}$ such that one has the commutation relations

$$[P^\mu, P^\nu] = 0$$

$$\frac{1}{i}[J^{\mu\nu}, P^\lambda] = g^{\mu\lambda}P^\nu - g^{\nu\lambda}P^\mu$$

$$\frac{1}{i}[J^{\mu\nu}, J^{\kappa\lambda}] = g^{\nu\lambda}J^{\mu\kappa} - g^{\mu\lambda}J^{\nu\kappa} - g^{\nu\kappa}J^{\mu\lambda} + g^{\mu\kappa}J^{\nu\lambda} \tag{8}$$

where

$$P^\mu = \int_\sigma d\sigma_\nu T^{\mu\nu}$$

$$J^{\mu\nu} = \int_\sigma d\sigma_\lambda [x^\mu T^{\lambda\nu} - x^\nu T^{\lambda\mu}]$$

and $T^{\mu\nu}$ is the energy momentum tensor. Since this criterion has virtually universal acceptance, we shall freely adopt it here as the condition for determining the covariance of the theory described by (2). In analogy to the well known Dirac-Schwinger covariance condition[2] appropriate to conventional coordinates one can readily derive the result[3]

$$[T^{++}(x), T^{+-}(x')] = i[\partial_- T^{+-}(x)]\delta(x^- - x^{-\prime}) \tag{9}$$

on the energy momentum tensor in two dimensional light cone coordinates.

By straightforward calculation one finds for the theory described by (2)

$$T^{\mu\nu} = g^{\mu\nu}L - \frac{i}{4}\psi\beta(\gamma^\mu\partial^\nu + \gamma^\nu\partial^\mu)\psi.$$

Application of the action principle yields the only independent commutator of the theory

$$\{\psi_1(x), \psi_1(x')\}\delta(x^+ - x^{+\prime}) = \delta(x-x')\frac{1}{\sqrt{2}}$$

with the commutators of ψ_2 to be determined from the constraint equation

$$\psi_2(x) = -\frac{m}{2\sqrt{2}}\int_{-\infty}^{\infty} \varepsilon(x^- - x^{-\prime})\psi_1(x')dx^- \tag{10}$$

obtained from the solution of Eq. (4). Direct calculation is found to yield an additional term on the right hand side of (9) of the form[4]

$$- \frac{m}{8} \partial_- \partial'_- [\psi_1(x)\varepsilon(x^- - x^{-'})\psi_2(x')]. \tag{11}$$

Although the term (11) is not compatible with (9), it is not immediately clear whether it will yield a contribution to (8) upon integration. Thus one calculates

$$[T^{+-}(x), J^{+-}] = -i(x^+\partial^- - x^-\partial^+)T^{+-}(x)$$
$$+ \frac{m^2}{8\sqrt{2}} \lim_{L\to\infty} L(\psi_1(L) - \psi_1(-L))\psi_1(x)$$

and uses the fact that the two point function of the free field behaves as $(x^- - x^{-'})$ for $x^+ = x^{+'}$ to infer that the unwanted term cannot be made to vanish and that the Poincaré algebra consequently cannot be realized in the case $m \neq 0$.

If one had approached the covariance problem not from the viewpoint of satisfying (8) but rather with the intent of showing that ψ transforms covariantly, identical results would have been obtained. In this case one finds that ψ_1 transforms precisely as it should but that the dependent field ψ_2 given by (10) transforms covariantly only for m=0. The point is that just as in the calculation of the commutators of the Poincaré generators, the long range behavior of the integrand leads to the breakdown of covariance.

While you may be willing to grant the validity of this formal breakdown of covariance as I have presented it, I can imagine that you remain at least partially unconvinced because I have not shown how calculations of amplitudes could fail to have the correct transformation properties. Let me therefore formulate and answer this objection.

It is well known that one can use the device of external sources to write down expressions for all matrix elements of a given theory. If all such matrix elements are defined in terms of variational derivatives with respect to such sources, then any amplitude can be calculated in terms of a covariant set of operations on the free field propagators. Except then for certain subtleties having to do with definitions of current operators (to which we shall shortly return) it would appear that S-matrix amplitudes must be covariant if the free propagators are. This way of expressing the problem indeed suggests that the free field propagator cannot be the same in light cone coordinates as in conventional coordinates if we are to obtain results compatible with the operator formalism. Furthermore the nonexistence of a covariant propagator must appear in the two dimensional spin one-half case precisely at the point at which mass is introduced if a precise correspondence is to be obtained.

Let us now demonstrate this remarkable result[5]. Defining the propagator as

$$G(x-x') = \frac{\delta}{\delta\eta(x')} <0|\psi(x)|0>\Big|_{\eta=0}$$

where we have made the replacement

$$L \longrightarrow L + \frac{1}{2}[\psi,\beta\eta]$$

in (2), one obtains

$$G(x-x')=i\varepsilon(x^{+}-x^{+'})<0|(\psi(x)\psi(x')\beta)_{+}|0>+<0|\frac{\delta\psi(x)}{\delta\eta(x')}|0>$$

or

$$G(x-x')=i\varepsilon(x^{+}-x^{+'})<0|(\psi(x)\psi(x')\beta)_{+}|0>+\frac{i}{4}\alpha^{-}\beta\delta(x^{+}-x^{+'})\varepsilon(x^{-}-x^{-'}). \tag{12}$$

Considering first the case $m=0$ one finds that G satisfies the equation

$$\gamma^{\mu}\frac{1}{i}\partial_{\mu}G(x-x') = \delta(x-x'). \tag{13}$$

Although one is tempted to infer from (13) that

$$G(p) = -\frac{\gamma p}{p^{2}-i\varepsilon}, \tag{14}$$

it is easily seen that this structure is incorrect for the case of $G_{21}(p)$. Thus from (13) it would follow that

$$G_{21}(p) = -\frac{i\sqrt{2}\,p^{-}}{p^{2}-i\varepsilon} = -\frac{i}{\sqrt{2}}\,\frac{1}{p^{+}-i\varepsilon(p^{-})\varepsilon}$$

$$= -\frac{i}{\sqrt{2}}\,\frac{P}{p^{+}} + \frac{\pi}{\sqrt{2}}\,\varepsilon(p^{-})\delta(p^{+})$$

where the principal value term is easily seen to correspond to the contact term in (12). Since the term $\varepsilon(p^{-})\delta(p^{+})$ must consequently be identified with the time ordered product of two ψ_2 fields, we immediately infer from the vanishing of ψ_2 the fact that (14) is not the solution of Eq. (13). More careful calculation of $G_{21}(p)$ shows that the correct result is

$$G\beta = \frac{\alpha^{+}p^{+}}{p^{2}-i\varepsilon} + \alpha^{-}\frac{P}{2p^{+}}$$

or

$$G(p) = -\frac{\gamma p}{p^2 - i\varepsilon} + \frac{i\pi}{2}\gamma^+\varepsilon(p^-)\delta(p^+). \tag{15}$$

Thus there exists in the massless case a well defined propagator which while differing from the conventional one is fully covariant by virtue of the fact that two-vectors merely scale (without mixing of components) under Lorentz transformations.

While it may seem extraordinary that Eq. (13) yields different results in two different quantization schemes, it turns out not to be difficult to account for this fact. In conventional coordinates both components of ψ are independent dynamical variables and the usual causal boundary conditions ($p^2 \to p^2 - i\varepsilon$) are indeed appropriate. If, however, one uses light cone coordinates the equation for G_{21} becomes

$$\sqrt{2}\,\partial_- G_{21} = \delta(x-x') \tag{16}$$

and since there is no "time" derivative in (16) the specification of causal boundary conditions is irrelevent to solving that equation. However, by giving the usual symmetrical boundary conditions in x^- one obtains in an unambiguous fashion the result of Eq. (15).

The extension to the massive case is now readily carried out and merely requires that one solve the equation

$$(\gamma p + m)G(p) = 1. \tag{17}$$

Calling the zero mass propagator $G_0(p)$, the massive result is formally given by the series

$$G(p) = G_o(p) \sum_{n=o}^{\infty} [-mG_o(p)]^n$$

which by use of (15) readily becomes

$$G(p) = (m - \gamma p)\frac{1}{p^2 + m^2 - i\varepsilon} + i\pi\gamma^+ |p^-| [\frac{p^+\delta(p^+)}{p^2 + m^2}]. \tag{18}$$

Although the last term in this equation formally vanishes for the case $m \neq 0$, it is not difficult to establish that the remaining term in (18) is not an acceptable solution of (17). This follows from the observation that if one subtracts from G_{21} the term proportional to $\delta(x^+ - x^{+\prime})$, there remains the time ordered product of two ψ_2 fields. However, from (10) it follows that for (18) to be an acceptable solution the time ordered product of two ψ_2 fields must vanish at least as fast as m^2 for small m. By direct calculation one finds instead that it is proportional to $(x^+)^{-1}$ in that limit, thereby establishing the contradiction. It is worth noting that the additional term in Eq. (15) for the massless Green's function serves precisely to cancel this unwanted (and inadmissible) contribution to $G_{21}(p)$ but is [as seen from Eq. (18)] incompatible with the presence of a mass term.

IV. UNACCEPTABLE ANOMALIES

Thus far we have limited consideration to the case of a free spin one-half field. Although the defects of that theory make it unacceptable as an ingredient in a theory of quark binding it is of at least some interest to investigate the problems which occur even if one overlooks (!) the lack of covariance. In particular we shall discuss what is conveniently referred to as the 't Hooft model[6] of quark binding as described by the Lagrangian

$$L = \frac{i}{2}\,\psi\alpha^\mu(\partial_\mu - igT_a A^a_\mu)\psi - \frac{m}{2}\,\psi\beta\psi + \frac{1}{4}\,F_a^{\mu\nu}F^a_{\mu\nu} - \frac{1}{2}\,F_a^{\mu\nu}(\partial_\mu A_\nu - \partial_\nu A_\mu + igA_\mu t^a A_\nu). \tag{19}$$

In writing (19) the matrices T_a and t_a are taken to be respectively the fundamental and adjoint representations of U(N), thereby guaranteeing the formal U(N) invariance of the theory.

Since the difficulty with which we shall be concerned can be displayed in the simpler Abelian (N=1) case we make that choice and also set m=0. Thus we use the charge matrix

$$q = \begin{pmatrix} 0 & -i \\ i & 0 \end{pmatrix}$$

rather than T_a to describe the charge or U(1) degree of freedom associated with the Hermitian field ψ. Since these simplifications imply the equation

$$\partial_\nu F^{\mu\nu} = gj^\mu \tag{20}$$

where

$$j^\mu = \frac{1}{2}\,\psi\alpha^\mu q\psi,$$

it is clear that just as in the usual Maxwell theory consistency requires that j^μ be conserved, i.e. that

$$\partial_\mu j^\mu = 0.$$

If one uses the result that $\psi_2=0$ (which condition is unmodified by the interaction) it is found using the light cone gauge

$$A^+ = A_- = 0$$

that

$$\partial_\mu j^\mu = \partial_+ j^+$$
$$= \lim_{\varepsilon \to 0} \sqrt{\frac{1}{2}}\, \partial_+[\psi_1(x+ \tfrac{1}{2}\varepsilon)q\psi_1(x- \tfrac{1}{2}\varepsilon)]$$

where ε is a vector in the spatial (x^-) direction. Straightforward calculation[7] then yields the result

$$\partial_\mu j^\mu = - \frac{e}{4\pi}\, \varepsilon^{\mu\nu} F_{\mu\nu}$$

where $\varepsilon^{\mu\nu}$ is the Levi-Civita tensor. Thus the 't Hooft model bears a basic inconsistency in its equations of motion which renders it unsuitable as a candidate for a quark binding model independent of the covariance problems already noted. Although anomalies have been known to occur in other two dimensional models (e.g., the Schwinger model) there exists the important difference that in the present example the anomaly is in the current coupled to the gauge field rather than to the axial current.

But surely (as has been suggested) the problem outlined here is merely an indication of a need for regularization or some other device. That such anomalies are in fact unavoidable defects of the 't Hooft model is best illustrated by a paper[8] by Yee and myself in which we give the only known solution of an interacting theory in light cone coordinates. In that work we have extended the theory described by (19) with N=1 and m=0 to include a boson mass so that (20) becomes

$$\partial_\nu F^{\mu\nu} = gj^\mu - \mu^2 A^\mu$$

and current nonconservation need not imply a contradiction. Although the calculations are lengthy the result is that the current is indeed not conserved and it is impossible to take the $\mu=0$ limit. Since this coincides with the conclusion obtained directly from the 't Hooft model and since the solution obtained in ref. 3 is unique, we are able to conclude that the anomaly is real and inescapable.

V. CONCLUSION

In this talk I have shown that the massive spin one-half field is noncovariant in two dimensional light cone coordinates. One naturally is tempted to ask whether the conditions on spin and dimensionality can in fact be removed. It is almost trivial to show[5] that spin one-half is also noncovariant in four dimentions. The extension to zero spin is somewhat more difficult because there it appears that upon doing the explicit calculations the energy momentum tensor transforms covariantly even though the spin zero field does not[8].

Since, however, the case of the spin one-half field is an absolute necessity if one is to build a world containing fermions, it seems safe to conclude that light cone quantization cannot be useful in the quark binding problem as currently conceived. Because of this it would seem useful to suggest here that further work on light cone quantization be focused solely upon the questions of consistency as discussed here rather than on questionable applications to model building.

REFERENCES

1. The incorrect result (7) has been derived, for example, by Y. Frishman in a CERN preprint.

2. J. Schwinger, Phys. Rev. 127, 324 (1962); P.A.M. Dirac, Rev. Mod. Phys. 34, 592 (1962).

3. C. R. Hagen and J. H. Yee, Phys. Rev. D13, 2789 (1976).

4. C. R. Hagen, Comments on a Paper by Callan, Coote and Gross, preprint UR-570.

5. C. R. Hagen and J. H. Yee, Failure of Light Cone Quantization in Spinor Field Theories, preprint, UR-582.

6. G. 't Hooft, Nucl. Physics B75, 461 (1974).

7. C. R. Hagen, Nucl. Physics B95, 477 (1975).

8. C. R. Hagen and J. H. Yee, in preparation.

More Freedom Now: An Explanation of Local Duality and Other Mysteries

H. DAVID POLITZER*
Lyman Laboratory
Harvard University
Cambridge, Massachusetts

ABSTRACT

Unambiguous predictions of asymptotic freedom for $1 \lesssim Q^2 \lesssim 15$ GeV^2 electroproduction are described and confirmed by the data. The logarithms of asymptotic freedom are observed; their scale determines the quark-gluon coupling to be small, but non-zero: $g^2(2\text{ GeV})/4\pi^2 \approx 0.15$. Local duality and a refined version of the Drell-Yan relation emerge from the field theory. A prediction for $R(= \sigma_L/\sigma_T)$ disagrees with current experiment.

Quantum chromodynamics is increasingly recognized as a serious candidate for the theory of the strong interactions. There are two obvious challenges: Does the particle spectrum of QCD agree with the observed hadrons? And do the detailed predictions for lepton-hadron scattering agree with experiment? I would like to report on progress on the second question, made in collaboration with Howard Georgi and Alvaro De Rújula[1].

Preliminary observations of non-scaling in high Q^2 μ-p scattering[2,3] are in qualitative agreement with the predictions of asymptotic freedom[4]. More conclusive confirmation of the theory requires higher precision data. Such data exists for electroproduction at SLAC[5]. However, in previous analyses[6] the theory was ambiguous for such low Q^2's. Only asymptotic ($Q^2 \to \infty$) predictions were made, and no way of choosing which scaling variable to use, e.g. x, x', x_W, ..., was given. Yet, it is known that virtually all non-scaling effects at SLAC can be absorbed into an astutely chosen variable. Furthermore, scaling extends down to Q^2 of 1 GeV^2, which is clearly not asymptotic relative to m_p^2.

---

*Junior Fellow, Harvard Society of Fellows.

Recently, Georgi and I[7] have shown that the operator product analysis determines precisely what variable to use when measuring the interaction corrections to the impulse approximation. We have subsequently done a phenomenological analysis of electroproduction down to 1 GeV^2.[1,8,20] I will first describe how the ambiguities of low Q^2 predictions are eliminated. This will involve a review of the logarithmic scaling violations; a derivation of the "right" scaling variable; a discussion of the approximations; and a presentation of predictions and fits.

I will then explain how to reconcile these "scaling" predictions with the observed resonances. Herein lies a tale: the local duality of Bloom and Gilman[9]. The unrefined predictions of the theory come in the form of moments of structure functions. I will first discuss how the resonances affect moments experimentally. We will then see that the theoretical moment dependence of the corrections to the leading "scaling" predictions are precisely what is needed to allow for the existence of resonances and to derive local duality. Local duality in this particular form is extremely successful. I will exhibit the prediction for the proton itself, i.e., the prediction of the elastic, magnetic form factor, $G_M(Q^2)$, from inelastic data. Our methods are even a phenomenological improvement in that the absolute magnitude of $G_M(Q^2)$ is successfully predicted, as well as the Q^2 dependence[10].

Finally, I will resolve the mystery of local duality and scalar targets -- perhaps you already knew -- and then give the predictions for $R(=\sigma_L/\sigma_T)$. It is impossible to fit the existent, crude data and retain any degree of sanity. An improvement of the experiment is therefore essential.

I. THE LOGS OF ASYMPTOTIC FREEDOM

Early works on asymptotic freedom[4,6,11] derived predictions like

$$\int \nu W_2 \, x^{n-2} \, dx \xrightarrow{Q^2 \to \infty} A_n [\log Q^2]^{-d_n}, \qquad (1)$$

$n \geq 2$, even. The d_n are positive and computed; the A_n must be fit from experiment. The obvious questions are: when does this behavior set in? What sets the scale of Q^2?[12] How big are the corrections? Is the relevant variable really x or some x'? I'll suppress for the time being a further ambiguity arising from the phenomenon of operator mixing, whose net effect is to make d_n and A_n matrices, and return to this later.

The derivation of Eq. (1) begins with the operator expansion of the forward Compton amplitude, whose discontinuity is related to the electroproduction structure functions[13]:

$$\int e^{iqx}\, dx \, \langle P|iT(J(x)J(o))|P\rangle = \sum_{n,\alpha} c_{n,\alpha}(q)\langle P|O^{n,\alpha}|P\rangle. \quad (2)$$

Let n be the spin of the local operator $O^{n,\alpha}$, and α runs over all possible other labels. The c(q) are computable for all n,α using renormalization group improved perturbation theory in $g^2(Q^2)$. Asymptotic freedom implies that the c(q) behave as powers of Q^2, given by dimensional analysis, times calculable logarithms. For a given n, the operator of lowest mass dimension will have the leading coefficient as $Q^2\to\infty$. Asymptotically, the n-2th moment is controlled by operators of spin n; so the d_n in Eq. (1) comes from the asymptotic behavior of the coefficient of the leading operator.

II. ξ SCALING

To make a testable prediction, we must follow the theory down from $Q^2=\infty$ to some more accessible value. There are two tasks that must be accomplished to make a theory for finite Q^2. We must first determine the contribution to νW_2 of a single $O^{n,\alpha}$, if its $c_{n,\alpha}(q)$ is known. Previously, this was done ignoring effects of order Q^2/ν^2. The second task involves estimating the size of the matrix elements of the $O^{n,\alpha}$ for different α. At infinite Q^2, which operators dominate is determined purely by the relative sizes of the calculable coefficient functions, c(q), because the $\langle P|O^{n,\alpha}|P\rangle$ are Q^2-independent. However, for finite Q^2 we must know the relative sizes of the products $c_{n,\alpha}(q) \times \langle P|O^{n,\alpha}|P\rangle$.

Dependence on Q^2/ν^2 is ignored in deriving Eq. (1). One can ask, though, what the contribution of a single $O^{n,\alpha}$ is to νW_2 without making such approximations. The key observation is that the matrix element of an operator of a given even spin has a fixed tensor structure[14,7]. This brings in an explicit target mass dependence. For instance for n=2,

$$\langle P|O^{2,\alpha}|P\rangle = A_2^{\alpha}[P^{\mu}P^{\nu} - \tfrac{1}{4}\, g^{\mu\nu}m_p^2]. \quad (3)$$

There is a similar expression for any n.

If we assume for the moment that the coefficient functions c(q) are those of a free, massless quark theory, then there is a single operator for each n, and its c(q) is essentially

$(Q^2)^{-n} q^{\mu_1} \ldots q^{\mu_n}$. Contraction with the generalizations of Eq.(3) leads to scaling in a new variable ξ [7] where

$$\xi = \frac{2x}{1+\sqrt{1+4x^2 m_P^2/Q^2}} = \frac{Q^2}{2m_P\nu} \frac{2}{1+\sqrt{1+Q^2/\nu^2}} . \tag{4}$$

It is perhaps inappropriate to say that ξ summarizes the target mass corrections. Rather it gives the Q^2/ν^2 effects, and the target mass just creeps into the definitions of W_1 and W_2 and the range of ξ.

ξ-scaling can be combined with the logarithmic dependence of the coefficient functions, c(q). This will be done shortly. An important question is under what circumstances are the ξ corrections significant? And if they are significant, is it sufficient to keep only the leading operators (those that occur in the free field expansion)?

Since $\xi=x+O(m_P^2/Q^2)$, using ξ will only be an improvement for $Q^2 \sim O(m_P^2)$. On the other hand, to use asymptotic freedom in computing Q^2 dependence, we must have $g(Q^2)$ small enough to be an expansion parameter. Hence for ξ to be of any practical value, $g(m_P^2)$ must be small.

If the quark gluon coupling is indeed that small, there is a systematic expansion procedure in which keeping only the leading operators of free field theory is the first approximation. I have two different but related arguments.

The first argument to show that the leading behavior comes from the operators of the free field operator product expansion is a refinement of the original twist analysis. "Twist" is defined as "dimension minus spin" so that asymptotically only the operators of minimum twist contribute to Eq. (1). These are in fact the operators of the free field expansion. If $g^2(Q^2)$ is small even down to $Q^2 \approx m_P^2$, the natural or fundamental hadronic mass scale must be yet smaller. That fundamental mass, call it M_o is likely more on the scale of Λ where

$$g^2(Q^2)/4\pi \approx \frac{12\pi}{27} \frac{1}{\log Q^2/\Lambda^2} . \tag{5}$$

M_o may be defined by the condition that $g^2(M_o)/4\pi=1$ and is probably associated with the phenomenological parameters of the "transverse momentum cut off" or the size of hadrons. In estimating the target matrix elements of local operators, perturbative ideas are clearly useless if we use M_o as the scale to define (or renormalize) g. However, dimensional analysis estimates are likely to be good for these matrix elements (since

$g^2(M_o)/4\pi=1$), using powers of M_o to give the right dimension. That is to say that we reasonably expect the effects of higher twist operators to be suppressed by powers of M_o^2/Q^2 in electroproduction. This is because higher twist operators measure properties like the transverse momentum distribution, which is of scale M_o.

An alternative argument[7] leads to the same conclusion. We start with an idea first developed to study heavy hadrons[15]. If we renormalize g at the mass of the heavy hadron, certain <u>inclusive</u> properties may be calculable perturbatively in that coupling. We can apply this idea to the proton. If the matrix elements of the leading twist operators are defined to be of order unity for each spin, then the next to leading twist four operators (after using the interacting quark equations of motion) are of order $g^2(m_P^2)\ m_P^2/Q^2$. This agrees qualitatively with the first argument if in fact $g^2(m_P^2)/4\pi\times m_P^2 \sim M_o^2$. We have deduced $g^2(m_P^2)/4\pi$ from a fit to electroproduction data and find this relation to be satisfied.

I claim it is an excellent approximation to treat the quarks as massless when computing coefficient functions. In electroproduction, virtually all quarks involved are light (i.e. u and d quarks), except perhaps a small fraction for $\xi<0.1$. We have argued elsewhere[7] that the effective light quark mass as it enters the coefficient functions, $c_{n,\alpha}(q)$, is very small ($\sim$20 MeV) for any Q^2 such that $g^2(Q^2)$ is small.

III. LOGARITHMIC CORRECTIONS TO ξ-SCALING

If only the leading operator of twist-2 is important for each spin (and only then can we get approximate scaling as a prediction rather than an accident) then the ξ-scaling and anomalous logarithmic dimension analyses can be combined. We find for instance[7]

$$\int \nu W_2 x^{n-2} dx = \sum_{j=0}^{\infty} \left(\frac{m_P^2}{Q^2}\right)^j \frac{n+j}{j!(n-2)!} \frac{A_{n+2j}(Q^2)}{(n+2j)(n+2j-1)} \tag{6}$$

where

$$A_m(Q^2) = A_m[a_o+a_1 g^2(Q^2) + \ldots][\log Q^2/\Lambda^2]^{-d_m} . \tag{7}$$

To obtain a formula for νW_2 itself, we can perform a formal moment inversion by introducing a function $F(y,Q^2)$ whose moments are $A_n(Q^2)$, i.e.

$$A_n(Q^2) = \int_0^1 dy\, y^n\, F(y,Q^2). \qquad (8)$$

The properties of $F(y,Q^2)$ given the d_n have been studied previously[16,17]. With the formal inversion I add the caution that it is to be trusted only in as much as the moment statements are accurate. We will see that this translates into a question of the degree of locality of local duality.

The inverted predictions are

$$\nu W_2(Q^2,x)/m_P = \frac{x^2}{(1+4x^2m_P^2/Q^2)^{3/2}} F(\xi,Q^2) + 6\frac{m_P^2}{Q^2}\frac{x^3}{(1+4x^2m_P^2/Q^2)^2}\int_\xi^1 d\xi' F(\xi',Q^2) + 12\frac{m_P^4}{Q^4}\frac{x^4}{(1+4x^2m_P^2/Q^2)^{5/2}}\int_\xi^1 d\xi'\int_{\xi'}^1 d\xi'' F(\xi'',Q^2) \qquad (9a)$$

$$W_1(Q^2,x) = \frac{x}{2\sqrt{1+4m_P^2/Q^2}} F(\xi) + \frac{m_P^2}{Q^2}\frac{x^2}{(1+4x^2m_P^2/Q^2)}\int_\xi^1 d\xi' F(\xi',Q^2)$$

$$+2\frac{m_P^4}{Q^4}\frac{x^3}{(1+4x^2m_P^2/Q^2)^{3/2}}\int_\xi^1 d\xi'\int_{\xi'}^1 d\xi'' F(\xi'',Q^2). \qquad (9b)$$

These results are slightly simpler in terms of W_L and W_T:

$$2x(2W_T-W_L) = 6xW_1-(1+4x^2m_P^2/Q^2)\nu W_2/m_P = \frac{2x^2}{\sqrt{1+4x^2m_P^2/Q^2}} F(\xi,Q^2), \qquad (9c)$$

$$2xW_L = (1+4x^2m_P^2/Q^2)\nu W_2/m_P-2xW_1 = 4\left\{\frac{m_P^2}{Q^2}\frac{x^3}{1+4x^2m_P^2/Q^2}\int_\xi^1 d\xi' F(\xi',Q^2)\right.$$

$$+2\,\frac{m_P^4}{Q^4}\,\frac{x^4}{(1+4x^2m_P^2/Q^2)^{3/2}}\int_\xi^1 d\xi'\int_{\xi'}^1 d\xi''F(\xi'',Q^2)\Big\}. \tag{9d}$$

IV. OPERATOR MIXING AND GLUONS

Predictions like Eq. (1) in fact only exist for difference functions like $\nu W_2^{\text{proton-neutron}}$. The asymptotic n^{th} moment of νW_2^{proton} is governed by three operators of spin n, dimension n+2. Consequently, for each n there are three independent constants, i.e. $A_n^{(1)}$, $A_n^{(2)}$ and $A_n^{(3)}$ (and three different exponents). The measurement of νW_2 at a single Q^2 does not determine all three and therefore is not sufficient to predict the subsequent Q^2 evolution.

The A_n are proportional to the proton matrix elements of the various operators and contain the information of the proton wave function. Computing the A_n is a genuine strong interaction problem and is beyond our present capacity. Known methods are only sufficient to predict the Q^2 evolution of electroproduction.

Two hypotheses that reduce the three possible powers to one per n come to mind. First, for each n there is a leading operator (or linear combination) whose exponent of $\log Q^2$ is the least negative. At $Q^2 \approx \infty$ this operator will be more important than the other two. This suggests a prescription for finite Q^2, i.e., keeping only the leading operator. While this agrees crudely for n=2 with the experimental determination of the A's, I think the argument is suspect. If, for instance the three A_n were comparable at some low Q^2, then, by the time Q^2 was sufficiently large that the factors $[\log Q^2]^{-d_n}$ were significantly different, even the leading one would be small compared to what it was at low Q^2; that is to say that the whole structure function would be concentrated at $\xi \approx 0$. While this will inevitably happen at super asymptotic Q^2 ($\log\log Q^2/\Lambda^2 \gg 1$), it is not an accurate description of present physics.

The second approach asks: what are reasonable expectations for the gluon distribution? The matrix elements of the three operators, i.e. the A_n, are essentially the moments of three parton distribution functions: isoscalar and isovector quark distributions and the gluon distribution. It is known from experiment that the areas (n=2) of the isoscalar and gluon distributions are roughly equal. In parton language, this is the statement that the quarks carry one half the momentum in the infinite momentum frame. However, the gluon distribution in the proton is likely concentrated at small ξ. This results from a picture in which there are no "valence" gluons but rather an infinite number of soft ones responsible for quark binding. The quark distribution arises principally from the three valence quarks -- which in free field theory give $\nu W_2 \propto \delta(\xi-1/3)$-- whose momenta are smeared out by the fact of spatial confinement. Hence the quark distribution peaks just below 1/3 and has a non-neglibible tail. In contrast, the only gluons in the proton are presumably the gluons of the gauge string, which contains an infinite number of them. So while half of the momentum resides in the strings, it is shared (roughly equally) among an infinite number of gluons, making their distribution function concentrated at $\xi=0$.

Furthermore, the gluons only effect electroproduction by making transitions to quark-antiquark pairs, and there are small upper bounds for the amount of antiquarks present at $\xi \gtrsim 0.1$. So for $\xi \gtrsim 0.2$, the gluon distribution is negligible compared to the quark distribution. In terms of moments, this means that for $n \geqslant 4$ the gluon operator matrix elements are negligible.

Note that the two above arguments actually agree for $n>4$ because the leading linear combination of operators is predominantly the quark operator for $n>4$. These estimates are also consistent with the known pattern of operator mixing. The presence of quarks induces in perturbation theory a gluon distribution, but that induced distribution is concentrated at much smaller ξ than the original quark distribution.

If we ignore the effects of gluons mixing back with the quarks, then the two quark operators (isoscalar and isovector) effectively have the same logarithmic behavior and can be lumped together into a single term per n.

To perform the moment inversion problem for $F(\xi,Q^2)$ we could use the anomalous dimension formulas for the d_n of the pure quark operators. The result would only be reliable for $\xi \gtrsim 0.2$ because of our ignoring the gluon effects. We attempt a slight improvement on this procedure. We construct a new formula for d_n satisfying two criteria. It agrees with the perturbation theory calculation of the anomalous dimensions of the quark twist-2 operators for large n. Second, it connects smoothly in n with the expected behavior of νW_2 for n=2, which can be deduced including gluon effects.

V. THE VARIOUS APPROXIMATIONS

To arrive at well defined predictions like Eq. (9), we must make a variety of assumptions, approximations, and expansions. I'd like to collect them together and describe them briefly.

The coefficient functions c(q) are computed in an expansion in $g^2(Q^2)$. The general form is

$$c(q) = [a_0+a_1g^2(Q^2) + \cdots] \exp \int_{M^2}^{Q^2} \gamma(g^2(Q'^2))\frac{dQ'^2}{Q'^2} \tag{10}$$

The factor in the square brackets is simply the direct evaluation of the coefficient function in perturbation theory, using Q^2 as the renormalization scale and $g(Q^2)$ as the coupling constant. The effective coupling $g(Q^2)$ is computed by integrating the renormalization group β function, which itself is computed in perturbation theory. The γ in the exponential is the anomalous dimension of the renormalization group equation satisfied by c(q).

We[1] have computed the coefficient functions to the first non-trivial order, i.e., a_0 and a_1 of Eq. (10). This, together with the already known β and γ's gives the coefficient functions to $O(g^2)$.

To eliminate the ambiguity introduced by operator mixing we assume that the gluon distribution is negligible compared to the quark distribution for $\xi \gtrsim 0.2$ (or $n \gtrsim 4$). To improve the analysis to include low ξ or better justify the assumption where we use it is beyond us at present. One would need a theory of the shape of the gluon distribution. Deducing the gluon distribution from observations of scaling violations is probably beyond the accuracy of current experiments, but improving the bounds on the absence of anti-quarks for $\xi \gtrsim 0.1$ would make the argument more compelling.

The assumption that we need consider only the "free field" twist-2 operators introduces errors of order M_0^2/Q^2 or $g^2(m_p^2)/4\pi \times m_p^2/Q^2$. To improve this approximation, we would need to admit into our expansion a set of operators of twist 4. Their matrix elements would also have to be deduced from experiment. This cannot be done by measurements of a single Q^2 but must be deduced by extracting a contribution to νW_2 that falls like $1/Q^2$ in addition to the behavior predicted by logarithmically corrected ξ-scaling. This may be feasible using extremely precise experiments for $1 \lesssim Q^2 \lesssim 4$ GeV^2.

VI. A PARTON DESCRIPTION OF ξ-SCALING

I repeat here a parton description of ξ-scaling[7,18] to shed light on the approximations in another language. ξ-scaling (g=0) can be described as follows: The struck quark has initially a momentum P_I which is "on shell" $P_I^2 \approx m_q^2 \approx 0$. Its transverse momentum, $P_\perp$, ($P_\perp \cdot P \equiv 0$, $P_\perp \cdot q \equiv 0$) is zero: $P_\perp^2 \approx 0$. The final momentum, $P_F = P_I + q$, is also "on shell": $P_F^2 \approx m_q^2 \approx 0$. The operator product expansion says that the wave function is a function of the light cone variable $\xi = P_I^+/P^+$, where the + component is defined as the 0 plus the 3 component in a system where $P=(P^0,0,0,P^3)$, $q=(q^0,0,0,q^3)$ and $P_I=(P_I^0,P_\perp^1,P_\perp^2,P_I^3)$. From the above information it is a matter of algebra to derive Eq. (4).

The logarithmic corrections to ξ scaling can be viewed as the leading interaction corrections to the impulse approximation, which include the effects of radiating gluons during the scattering process. This softens the effective quark distribution with increasing Q^2.

Of the same order in g is a process whereby gluons contribute to the effective quark distribution by making transitions to quark-antiquark pairs. But if there are virtually no hard gluons ($\xi \gtrsim 0.2$), then the soft ones don't make hard quarks. Furthermore, there is no evidence for hard antiquarks. Hence we ignore this gluon process for $\xi \gtrsim 0.2$.

The leading effects of the interactions on the scattering process are logarithmic, characteristic of a renormalizable theory. The interactions also serve to confine the quarks and complicate our simple, one variable picture of the wavefunctions. But these effects introduce a fundamental scale, the confinement radius or the transverse momentum cut off. If we restrict ourselves to Q^2 of 1 GeV^2 and above, we argue that the confinement effects are of order $\langle P_\perp^2 \rangle / Q^2$ which is around 10% at Q^2= 1 GeV^2 and rapidly decreasing. To improve on our leading approximation we would have to allow the quark distribution functions to depend on $P_\perp^2$. In practice, it would be difficult to get a significant determination of the $P_\perp$ dependence from the available data on νW_2.

VII. FITS AND PREDICTIONS

Figure 1 shows recent data from SLAC[5] for νW_2 in the deep inelastic region. We have made four parameter fits using Eq.(9) as follows: we take $F(\xi, Q^2{=}15\ GeV^2)$ to be of the form

$$F(\xi,\ Q^2 = 15) = A\xi^{-B}(1-\xi)^C. \tag{11}$$

We fit A, B and C and Λ, which determines the size of $g(Q^2)$ through Eq. (5) and sets the scale of the logarithms. Our best overall fit is Λ=0.5 GeV, A=2.9, B=1.2, and C=3.5. It is

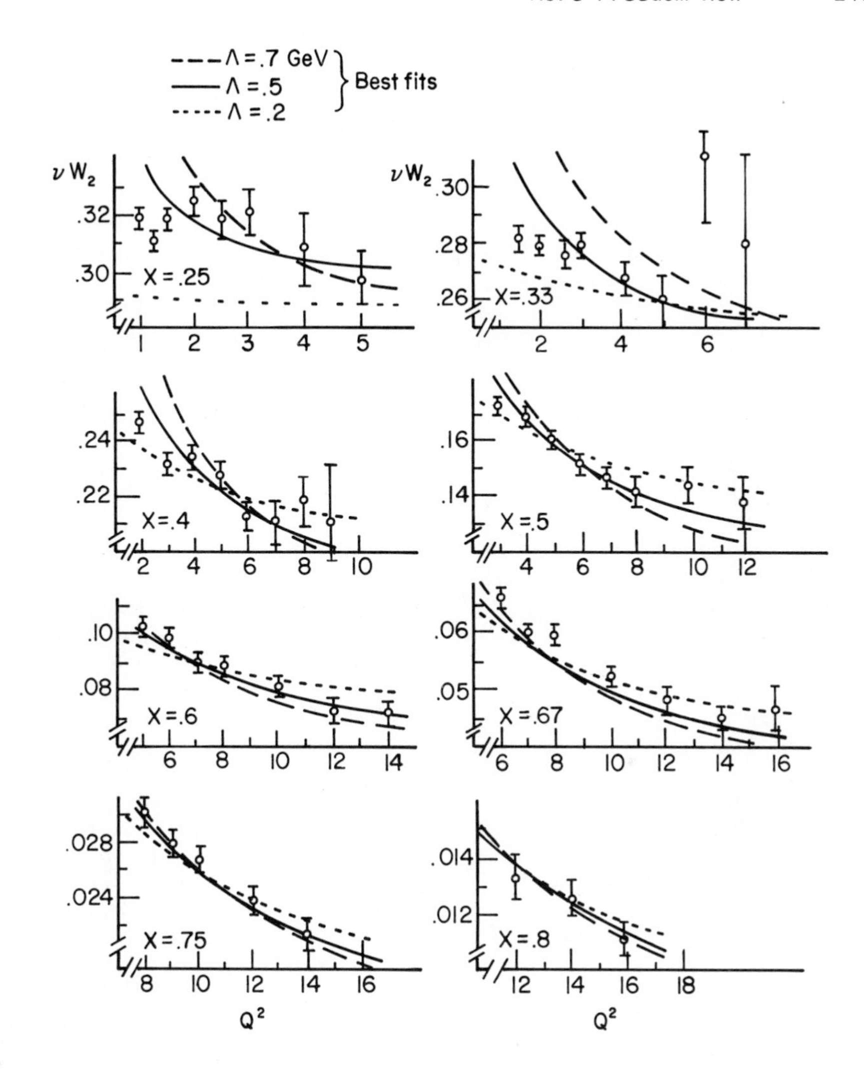

FIGURE 1. νW_2 as a function of Q^2 for different x; curves represent fits for different values of Λ; Λ=0.2 is too flat, Λ=0.7 is too steep.

certainly possible to change one parameter a little, compensate with another, and still go through the available data. However, if Λ gets too large or too small, no choice of A, B, and C is satisfactory. For example, the best fits with Λ=0.2 and Λ=0.7 are shown in Fig. 1.

Figure 2 shows the Q^2 evolution of $\xi^2 F(\xi,Q^2)$ for Λ= 0.5 GeV. Note that the fractional changes are large as $\xi \to 1$. And note also that if indeed $\Lambda \sim 0.5$ GeV, then the range of Q^2 available at SLAC is ideal for studying the logarithmic violations of scaling, while for $30 \lesssim Q^2 \lesssim 100$ GeV^2, results on scaling violations are not likely to be statistically significant. Thus, the most interesting experimental program in this field is to fill in the Q^2-ν plot for $Q^2 \lesssim 20$ GeV^2.

VIII. LOCAL DUALITY

Figure 3 shows the data for νW_2 for Q^2=2,3 and 5 GeV^2,[5] just as examples. Plotted also is the extrapolation to low Q^2 of the non-resonant data of Fig. 2. There are two striking features to these curves: locally in ξ they bear no relation. Yet when averaged over the width of a resonance, they agree remarkably. The agreement over the whole range $1 \lesssim Q^2 \lesssim 15$ when integrated over the width of a resonance is about 5% for the heaviest prominent resonance, 7% for the next, and 10% for the lightest.

Figure 4 demonstrates the quality of local duality for the proton itself. The data points are measurements of $G_M(Q^2)$, the proton elastic magnetic form factor, which is proportional to the area under the elastic peak in νW_2. The dotted line is the prediction for $G_M(Q^2)$, using the non-resonant data of Fig. 1 and using our precise version of local duality in terms of moments (to be discussed below). Above Q^2 of 5 GeV^2, the proton position, $\xi_p = \xi(x=1,Q^2)$ is above ξ=.8, the highest ξ for which we have non-resonant data. Instead of extrapolating our non-resonant fit, we re-did the calculation using the whole structure function at 15 GeV^2 as input, including elastic and quasielastic cross sections. Hence the agreement at 15 GeV^2 is automatic. Over the whole range of available data, the discrepancy with data is less than 25%.

To see that this is indeed expected from the theoretical standpoint, we must consider the corrections to the leading behavior. This is most easily described in terms of moments.

First, how in fact do the resonances affect moments? A resonance at fixed mass m will reside at a value of ξ (call it ξ_m) which goes to 1 as Q^2 increases. For $Q^2 \gg m^2$, $\xi_m \approx m^2/Q^2$.

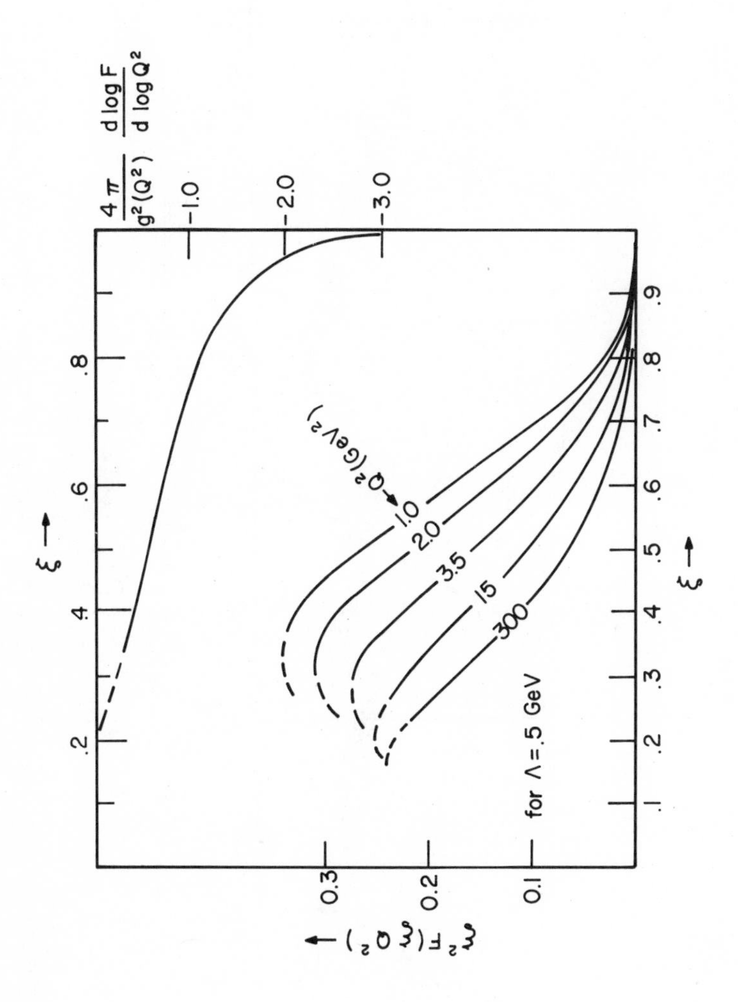

FIGURE 2. The Q^2 evolution of our fit to $\xi^2 F$ ($\approx \nu W_2$) for Λ=0.5.

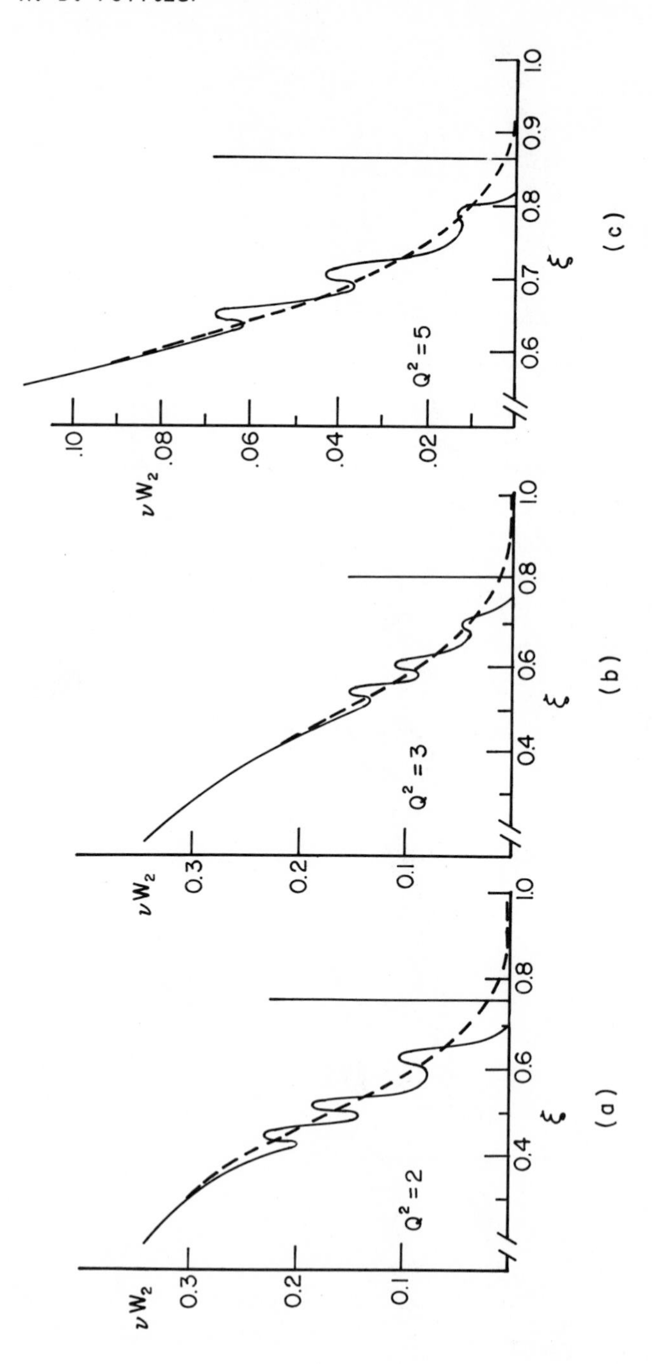

FIGURE 3. The resonances in νW_2 compared to the extrapolated non-resonant fit.

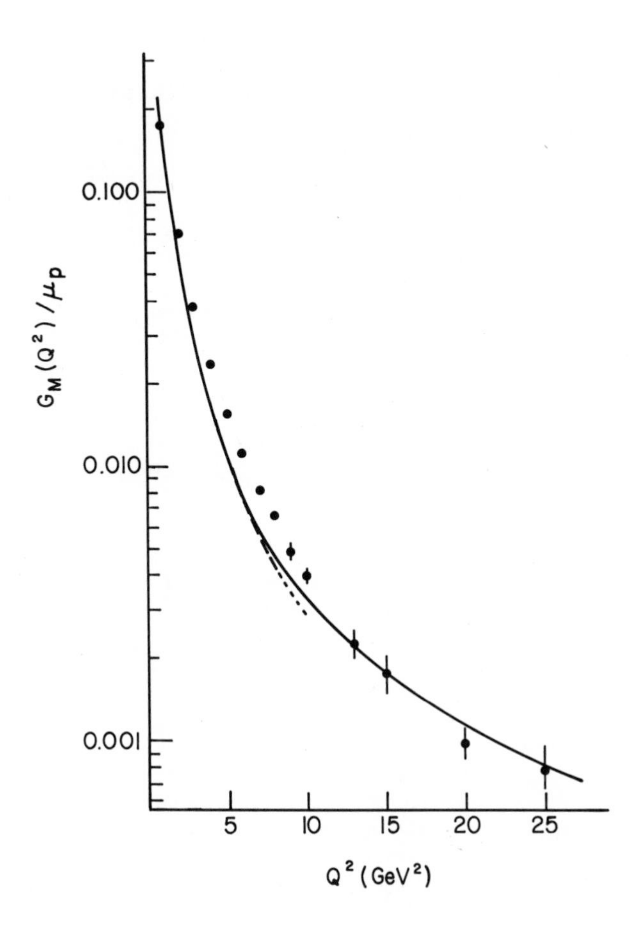

FIGURE 4. The magnetic form factor: data and prediction from improved local duality.

νW_2, which the resonances follow, is rapidly decreasing in ξ for $\xi>0.3$, something like $(1-\xi)^{3.5}$. So as Q^2 increases, the contribution of a resonance to a particular moment decreases as $\xi_m\to 1$. The contribution from the resonance eventually becomes negligible compared to the uncertainty in the whole moment arising from errors in the determination of νW_2 for smaller ξ. To study a resonance using moments, we must increase n, the power of ξ, with increasing Q^2. For a resonance to give a fixed fraction of the nth moment, n must increase like $n=KQ^2/m^2$. $K\sim 1.5$ will ensure that a resonance of mass m accounts for about half of the $n=KQ^2/m^2$ moment. Note also that for a given n (not too large, i.e. $n\lesssim 4Q^2/m^2$), the heavier resonances will make a greater contribution to the nth moment, and hence the area under them is better determined by that moment.

If moment predictions like Eq. (6) were true for arbitrarily high n then νW_2 would have to scale (up to logarithsm) locally in ξ. The characteristic wiggling of the resonances, with their characteristic m^2/Q^2 dependence would not be allowed. But, of course, Eq. (6) is only an approximation. We have included effects of $O(m_p^2/Q^2)$ but neglected effects of $O(\langle P_\perp^2\rangle/Q^2)$ in the form or higher twist operators. Furthermore, upon inspection,[1] these twist-4 operators have coefficient functions which have a factor of n, their spin, relative to the twist-2 operators. So the terms we have ignored are in fact of $O(n\langle P_\perp^2\rangle/Q^2)$! Higher twists have coefficients with increasing powers of n, i.e. $O([n\langle P_\perp^2\rangle/Q^2]^m)$. So for a given Q^2, predictions like Eq. (6) are limited to $n\lesssim N(Q^2)$ where $N(Q^2)$ increases linearly with Q^2. For $n>N(Q^2)$, perturbation theory is useless. Presumably, it is the same effects which prevent us from going to high n as bind the resonances themselves. We cannot really predict the constant of proportionality between $N(Q^2)$ and Q^2, but we can deduce it phenomenologically. We simply ask how high in n can we go before Eq. (6) breaks down because of the moving resonant bumps. Clearly $N(Q^2)$ is big enough to isolate a single resonance but not so big as to probe the shape of the resonance itself.

(In Ref. 1, we give a complete description of all the n dependent errors. We also explain how moment statements can be translated into smearing function statements. For example, with $N(Q^2)$ moments at our disposal, we can construct polynomials in ξ of degree N and predict scaling of νW_2 integrated with the polynomials. Since the N coefficients are up to us, we can sample νW_2 around any particular ξ_o with a width determined by N.)

IX. MESON TARGETS AND LOCAL DUALITY

There has long been a paradox associated with local duality and meson form factors. Imagine scattering an electron off a spin-0 hadron. The deep inelastic structure functions would presumably reflect the spin-1/2 nature of the fundamental charged constituents. They would satisfy the Callan-Gross relation, at least approximately. However, a spinless hadron has only one elastic form factor, and its contribution to the total electroproduction cross section is purely scalar, i.e., $W_1/\nu W_2 \approx 0$. Hence, a meson elastic form factor cannot be dual to the essentially transverse inelastic structure function. It has the wrong tensor structure or angular dependence. So why should there be a relation for the proton when there can't be one for the pion?

This dilemma is particularly acute for the present analysis. In the field-theoretic, operator product approach, no assumptions are made regarding the target hadron spin. All discussions of spin averaged inelastic structure functions apply equally well to spin-0 and spin-1/2 targets. Fortunately, the answer to this problem is simple; it even provides some new information on how local is local duality.

The key observation is that the transition form factors, which constitute the quasi-elastic bumps in the inelastic structure functions, have their own characteristic spin dependence. It is the sum of the elastic and transition form factors which is dual to the smooth inelastic structure function observed at higher Q^2. The meson form factor paradox arose only because we assumed that duality was sufficiently local to separate unequivocally a spin-0 elastic final state from its spin-1 hyperfine partner. Evidently, that is just impossible.

Let us discuss a simple model in which it is clear that the spin-0 to -1 transition form factor is precisely what is needed to make local duality as good for a spin-0 target as for a spin-1/2 target. Consider a charged, spin-1/2 parton bound to an electrically neutral, spin-1/2 object; the binding interaction is approximately spin-independent; hence there are two degenerate s-wave states, spin 0 and spin 1, whose spatial wavefunctions are the same. The electron scattering cross section will include a contribution from elastic scattering in which the charged parton spin is left unchanged and a contribution from the quasi-elastic transition to the spin-1 bound state, in which the parton spin is flipped. Clearly, there are no interference terms in the cross section between the amplitudes for these processes. Consider now a third degenerate state in which the charged parton is bound to a neutral spin-0 object in the same spatial, s-wave state, hence total spin 1/2. This bound state will have two elastic form factors: an electric form factor corresponding to no spin flip and a magnetic form factor for the spin flip transition. These are equal respectively to the elastic and transition form factors of the spin-0 state described above.

So in the limit of spin-independent binding forces -- presumably a fair zeroth approximation for quarks and gluons -- the situation is straightforward. Spin dependent forces not only induce a hyperfine splitting but also perturb the vector meson wave function relative to the scalar meson, and hence the two form factors will no longer be trivially related. Duality cannot be so local as to separate the elastic and quasi-elastic peaks. So if one wished to determine the elastic form factor of a spinless hadron from inelastic scattering data, the quasi-elastic cross section would have to be measured and explicitly subtracted out.

X. $R(=\sigma_L/\sigma_T)$

I have discussed W_2 but could do the same for W_1. It is more interesting to consider the combination $R=W_L/W_T=\sigma_L/\sigma_T$, which vanishes in a free theory of massless spin-1/2 partons and is known experimentally to be small. We can combine the asymptotic prediction for R[19] with the ξ-scaling behavior under the assumptions that the contributions of higher twist and gluon operators are small. Since this restricts us to $\xi>0.2$ anyway, we can approximate W_T as

$$W_T = A(Q^2)\xi(1-\xi)^{a(Q^2)}. \tag{12}$$

$a(Q^2)$ is a slowly decreasing function of Q^2 and is approximately 3-3½ over the range of interest. We find

$$R = \frac{1-\xi}{a+1}\left[4\,\frac{m_P^2}{Q^2}\,\frac{\xi^{a+3/a+2}}{1+\xi^2 m^2/Q^2} + \frac{g^2(Q^2)}{12\pi^2}\,\frac{1}{\xi^{a/a+2}}\right] \tag{13}$$

This is plotted in Fig. 5 for $\Lambda=.5$ GeV. The data points are from Ref. 5, and the errors are largely systematic and probably underestimated. The discrepancies are not alarming for small ξ, but for $x\gtrsim.67$ there's clear disagreement.

If we increased Λ, thus increasing $g(Q^2)$, we could fit the data for x=.67 and .75. However, it requires $\Lambda\sim3$ GeV to get $R\sim0.2$ in that range. Since we cannot claim to predict anything below $Q^2\sim2\Lambda^2$ (or more likely $Q^2\sim4\Lambda^2$), $\Lambda=3$ Gev would rule out any possible explanation of approximate scaling at SLAC. So either

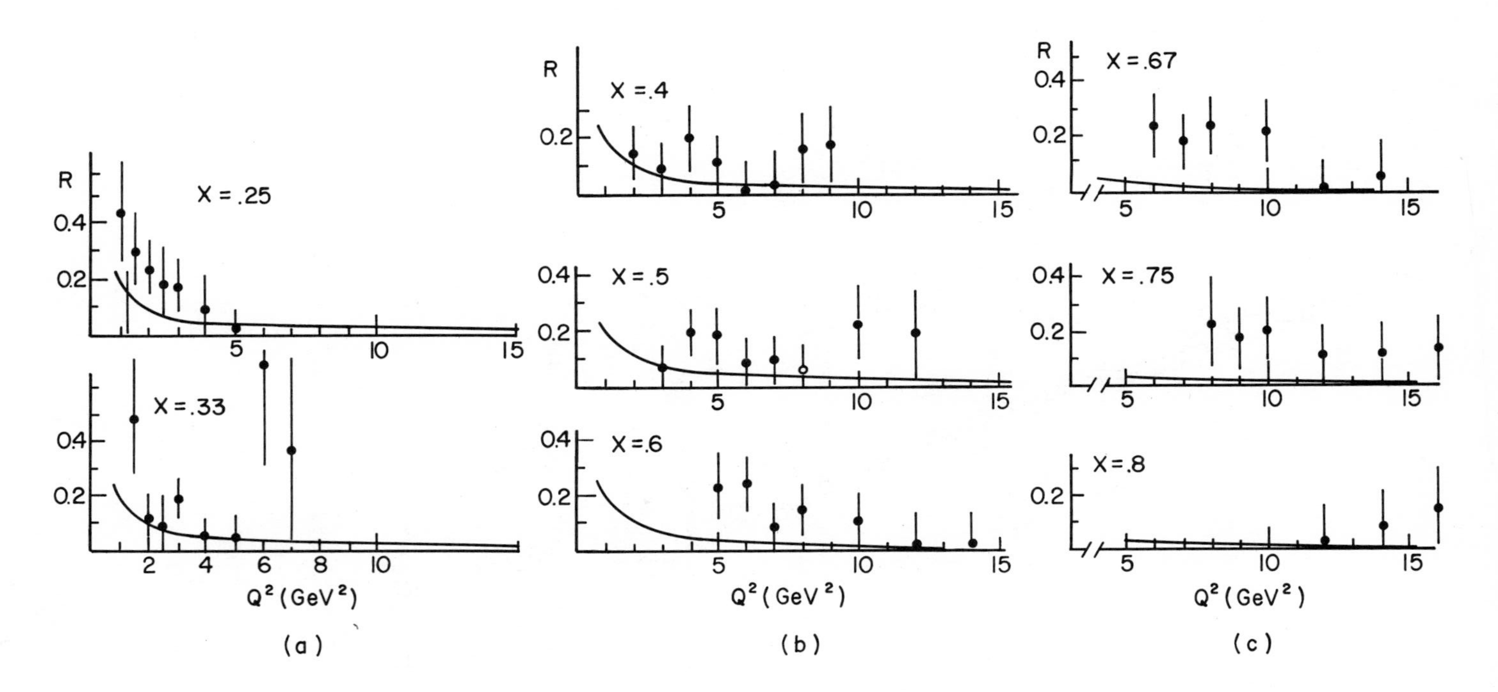

FIGURE 5. Data and predictions for $R(x,Q^2)$ ($= \sigma_L/\sigma_T$).

the theory is useless or the quoted measurements of R are wrong, e.g. 10 times too large at large x. So, difficult as the systematics make it, R remains a crucial quantity in testing theories of proton structure. Until the measurements of R can be improved, perhaps the theoretical prediction for R should be used when extracting νW_2 from unseparated data.

REFERENCES

1. A. De Rujula, H. Georgi and H. D. Politzer, "Demythification of Electroproduction Local Duality and Precocious Scaling," Harvard preprint, 6/76. Numerical results presented in this talk are preliminary so please refer to this ref. 1 for final numbers.
2. Y. Wanatabe et al., Phys. Rev. Lett. 35, 898 (1975); C. Chang et al., Phys. Rev. Lett. 35, 901 (1975).
3. H. Anderson et al., Phys. Rev. Lett. (to be published).
4. H. D. Politzer, Phys. Reports 14C, 129 (1974).
5. E. Riordan et al., SLAC-PUB-1634, August 1975. We are grateful to A. Bodek, D. Dubin and E. Riordan for providing a fit to their resonance data.
6. e.g. D. Gross and F. Wilczek, Phys. Rev. D8, 3633 (1973) and D9, 980 (1974).
7. H. Georgi and H. D. Politzer, Phys. Rev. Lett. 36, 1281 (1976) and "Freedom at Moderate Energies: ...," Harvard preprint, 2/76.
8. Ref. 7 also has important phenomenological implications for production of new quantum numbers in neutrino scattering; see ref. 7 and M. Barnett, Phys. Rev. Lett. 36, 1163 (1976).
9. E. Bloom and F. Gilman, Phys. Rev. D4, 2901 (1971).
10. S. Drell and T. M. Yan, Phys. Rev. Lett. 24, 181 (1970).
11. H. Georgi and H. D. Politzer, Phys. Rev. D9, 416 (1974).
12. H. D. Politzer, Phys. Rev. D9, 2174 (1974).
13. N. Christ, B. Hasslacher and A. Mueller, Phys. Rev. D6, 3543 (1972).
14. O. Nachtmann, Nucl. Phys. B63, 237 (1973).

15. T. Appelquist and H. D. Politzer, Phys. Rev. D5, 1404 (1975).

16. G. Parisi, Phys. Lett. 43B, 207 (1973).

17. A. De Rujula, H. Georgi and H. D. Politzer, Phys. Rev. D10, 2141 (1974).

18. P. Frampton, Phys. Rev. Lett. (to be published).

19. F. Wilczek, S. Treimen and A. Zee, Phys. Rev. D10, 2881 (1974).

20. V. Baluni and E. Eichten ("Precocious Scaling: Evidence for Field Theory," Inst. for Advanced Study preprint, 1976) have reached similar conclusions regarding low Q^2, non-resonant data and the effective coupling using a more general theoretical analysis.